Cira Perna

Marilena Sibillo

Mathematical and Statistical Methods in Insurance and Finance

Cira Perna (Editor)
Marilena Sibillo (Editor)

Mathematical and Statistical Methods in Insurance and Finance

Springer

Cira Perna
Dipartimento di Scienze Economiche e Statistiche
Università di Salerno, Italy

Marilena Sibillo
Dipartimento di Scienze Economiche e Statistiche
Università di Salerno, Italy

ISBN-13 978-88-470-5601-5 Springer Milan Berlin Heidelberg New York

Springer is a part of Springer Science+Business Media
springer.com

Softcover re-print of the Hardcover 1st edition 2008

Cover-Design: Simona Colombo, Milan
Typesetting with LaTeX: PTP-Berlin, Protago TeX-Production GmbH, Germany (www.ptp-berlin.eu)
Printing and Binding: Grafiche Porpora, Segrate (MI)

Preface

The MAF2006 Conference, organized at the University of Salerno, represents the prosecution of the first edition, held in 2004 also at the Campus of Fisciano, and was developed on the basis of cooperation between mathematicians and statisticians working in insurance and finance fields.

The idea arises from the belief that the interdisciplinary approach can improve research on these topics, and the proof of this is that interest in this guideline has evolved and been re-enforced.

The Conference aims at providing state of the art research in development, implementation and real word applications of statistical and mathematical models in actuarial and finance sciences, as well as for discussion of problems of national and international interest.

These considerations imply the strengthening of the involved methods and techniques towards the purpose, shared by an increasing part of the scientific community, of the integration between mathematics and statistics applied in finance and insurance fields.

The Conference was open to both academic and non-academic communities from universities, insurance companies and banks, and it was specifically designed to contribute to fostering the cooperation between practitioners and theoreticians in the field.

About 100 researchers attended the Conference and a total of 9 contributed sessions and 4 organized sessions, containing more than 50 communications, were accepted for presentation. The extended abstracts are available on `http://www.labeconomia.unisa.it/maf2006/`. Two prestigious keynote lecturers, delivered by prof. Giovanni Barone Adesi (University of Lugano) on *GARCH Options in Incomplete Markets* and by prof. Michael Wolf (University of Zurich) on *Multiple Testing Based on Generalized Error Rates with an Application to Hedge Fund Evaluation*, increased the scientific value of the meeting.

The collection published here gathers some of the papers presented at the conference MAF2006 and successively worked out to this aim. They cover a wide variety of subjects:

- *Mathematical Models for Insurance*
  Insurance Portfolio Risk Analysis, Solvency, Longevity Risk, Actuarial models, Management in Insurance Business, Stochastic models in Insurance.
- *Statistical Methods for Finance*
  Analysis of High Frequency Data, Data Mining, Nonparametric methods for the analysis of financial time series, Forecasting from Dynamic Phenomena, Artificial Neural Network, Multivariate Methods for the Analysis of Financial Markets.
- *Mathematical Tools in Finance*
  Stock Market Risk and Selection, Mathematical Models for Derivatives, Stochastic Models for Finance, Stochastic Optimization.

The papers follow in alphabetical order from the first author.

The quality of the papers is due to the authors and, in the name of the scientific and organizing committee of the conference MAF2006, we truly thank them all.

Moreover we thank the Faculty of Economics and the Department of Economics and Statistics of the University of Salerno for the opportunity they gave us to go ahead with this idea. We would like to express our gratitude to the members of the Scientific and Organizing Committee and to all the people who contributed to the success of the event. We are grateful for the kind effort in particular of the sponsors: Amases, Ambasciata di Svizzera a Roma, Associazione Costruttori Edili Salernitani, for making the meeting more comfortable and pleasant.

Finally, we truly thank the Department of Applied Mathematics and the Department of Statistics of the University of Venice for the strong cooperation in this initiative and for the involvement in organizing and hosting the next edition of the Conference, to be held in 2008 in Venice.

Fisciano, August 2007 *Cira Perna and Marilena Sibillo*

Contents

List of Contributors

Alessandra Amendola
University of Salerno
alamendola@unisa.it

Laura Attardi
University of Naples
attardi@unina.it

Anna Rita Bacinello
University of Trieste
bacinel@units.it

Diana Barro
University of Venice
d.barro@unive.it

Sergio Bianchi
University of Cassino
sbianchi@eco.unicas.it

Antonella Campana
University of Molise
campana@unimol.it

Elio Canestrelli
University of Venice
canestre@unive.it

Marta Cardin
University of Venice
mcardin@unive.it

Silvia Centanni
University of Verona,
silvia.centanni
@economia.univr.it

Rocco Roberto Cerchiara
University of Calabria
rocco.cerchiara@unical.it

Roy Cerqueti
University of Rome *La Sapienza*
roy.cerqueti@uniroma1.it

Pierangelo Ciurlia
University of Venice
ciurlia@unive.it

Rosa Cocozza
University of Naples *Federico II*
rosa.cocozza@unina.it

Cinzia Colapinto
University of Milan
cinzia.colapinto@unimi.it

Mariarosaria Coppola
University of Naples *Federico II*
m.coppola@unina.it

Marco Corazza
University of Venice
corazza@unive.it

Valeria D'Amato
University of Naples *Federico II*
valeriadamato@virgilio.it

Giovanni De Luca
University of Naples *Parthenope*
giovanni.deluca
@uniparthenope.it

Emilia Di Lorenzo
University of Naples *Federico II*
diloremi@unina.it

Marisa Faggini
University of Salerno
mfaggini@unisa.it

Paola Ferretti
University of Venice
ferretti@unive.it

Matteo Fini
University of Milan
matteo.fini@unimi.it

Francesco Giordano
University of Salerno
giordano@unisa.it

Giuseppe Giordano
University of Salerno
ggiordan@unisa.it

Luca Grilli
University of Foggia
l.grilli@unifg.it

Steven Haberman
City University, London, UK
S.Haberman@city.ac.uk

Davide La Torre
University of Milan
davide.latorre@unimi.it

Susanna Levantesi
University of Rome *La Sapienza*
susanna.levantesi@uniroma1.it

Francesco Lisi
University of Padova
francesco.lisi@unipd.it

Massimiliano Menzietti
University of Calabria
massimiliano.menzietti
@unical.it

Marco Minozzo
University of Perugia,
minozzo@stat.unipg.it

Martina Nardon
University of Venice
mnardon@unive.it

Marcella Niglio
University of Salerno
mniglio@unisa.it

Albina Orlando
Consiglio Nazionale delle Ricerche
a.orlando@na.iac.cnr.it

Edoardo Otranto
University of Sassari
eotranto@uniss.it

Graziella Pacelli
University of Ancona
g.pacelli@univpm.it

Maria Lucia Parrella
University of Salerno
mparrella@unisa.it

Paolo Pianca
University of Venice
pianca@unive.it

Augusto Pianese
University of Cassino
pianese@unicas.it

Massimiliano Politano
University of Naples *Federico II*
politano@unina.it

Giorgia Rivieccio
University of Naples *Parthenope*
giorgia.rivieccio
@uniparthenope.it

Giulia Rotundo
University of Tuscia, Viterbo
giulia.rotundo@uniroma1.it

Massimo Alfonso Russo
University of Foggia
m.russo@unifg.it

Maria Russolillo
University of Salerno
mrussolillo@unisa.it

Massimo Salzano
University of Salerno
salzano@unisa.it

Marilena Sibillo
University of Salerno
msibillo@unisa.it

Alessandro Trudda
University of Sassari
atrudda@uniss.it

Domenico Vistocco
University of Cassino
vistocco@unicas.it

Cosimo Vitale
University of Salerno
vitale@unina.it

Paola Zuccolotto
University of Brescia
zuk@eco.unibs.it

Scientific Committee

Cira Perna (Chair, University of Salerno),
Marilena Sibillo (Chair, University of Salerno),
Alessandra Amendola (University of Salerno),
Francesco Battaglia (University of Roma *La Sapienza*),
Elio Canestrelli (University of Venezia *Cà Foscari*),
Mariarosaria Coppola (University of Napoli *Federico II*),
Marco Corazza (University of Venezia *Cà Foscari*),
Alessandro of Lorenzo (University of Napoli *Federico II*),
Emilia of Lorenzo (University of Napoli *Federico II*),
Francesco Giordano (University of Salerno),
Michele La Rocca (University of Salerno),
Francesca Parpinel (University of Venezia *Cà Foscari*),
Paolo Pianca (University of Venezia *Cà Foscari*),
Claudio Pizzi (University of Venezia *Cà Foscari*),
Isabella Procidano (University of Venezia *Cà Foscari*),
Giuseppe Storti (University of Salerno),
Cosimo Vitale (University of Salerno)

Organizing Committee

Marcella Niglio (Coordinator, University of Salerno),
Diana Barro (University of Venezia),
Mariarosaria Coppola (University of Napoli *Federico II*),
Rosa Ferrentino (University of Salerno),
Stefania Funari (University of Venezia *Cà Foscari*),
Margherita Gerolimetto (University of Venezia *Cà Foscari*),
Maria Lucia Parrella (University of Salerno),
Maria Russolillo (University of Salerno)

Least Squares Predictors for Threshold Models: Properties and Forecast Evaluation

Alessandra Amendola, Marcella Niglio and Cosimo Vitale

Summary. The forecasts generation from models that belong to the threshold class is discussed. The main problems that arise when forecasts have to be computed from these models are presented and, in particular, least squares, plug-in and combined predictors are pointed out. The performance of the proposed predictors are investigated using simulated and empirical examples that give evidence in favor of the forecasts combination.

Key words: Threshold models; Forecasts generation; Forecast combination.

1 Introduction

Since their introduction at the end of the '70's, threshold models have been widely applied to study economic and financial time series.

The interest arisen from this class of models is even testified by the relevant number of variants proposed in literature with respect to the original one introduced in [Ton78]. Among them, in the present paper we give attention to the so-called *Self Exciting Threshold Autoregressive Moving Average* (SETARMA) models proposed in [Ton83], and recently revised in [ANVss], that are a direct generalization of the linear ARMA structure ([BJ76]). The original form, with k-regimes, of the SETARMA model of order $(k; p_1, \ldots, p_k; q_1, \ldots, q_k)$ takes form

$$X_t = \phi_0^{(i)} + \sum_{j=1}^{p_i} \phi_j^{(i)} X_{t-j} + \sum_{j=0}^{q_i} \theta_j^{(i)} e_{t-j} \qquad \text{with } e_t \sim WN(0, \sigma^2), \tag{1}$$

conditional on the threshold value $X_{t-d} \in R_i$, where $R_i = [r_{i-1}, r_i]$ forms a partition of the real line such that $-\infty = r_0 < r_1 < r_2 < \ldots < r_k = +\infty$, r_i are the threshold values, d is the threshold delay, p_i and q_i are non-negative integers.

The revised form of model (1) proposed in [ANVss], can be given as:

$$X_t = \sum_{i=1}^{k} [\phi_0^{(i)} + \phi_{p_i}^{(i)}(B) X_t^{(i)} + \theta_{q_i}^{(i)}(B) e_t^{(i)}] I(X_{t-d} \in R_i), \tag{2}$$

Cira Perna, Marilena Sibillo (Eds.): Mathematical and Statistical Methods in Insurance and Finance

where B is the backshift operator, the two polynomials $\phi_{p_i}^{(i)}(B) = \sum_{j=1}^{p_i} \phi_j^{(i)} B^j$ and $\theta_{q_i}^{(i)}(B) = 1 - \sum_{w=1}^{q_i} \theta_w^{(i)} B^w$ have no roots in common within each regime, $e_t^{(i)} = \sigma_i^2 e_t$, $\{e_t\}$ are i.i.d. random variables, with $E[e_t] = 0$ and $E[e_t^2] = 1$, for $i = 1, \ldots, k$ and $I(\cdot)$ is an indicator function.

The main variants that can be appreciated in model (2) are related to: the variance of the error component (that is left to change between the two regimes); the delayed values of X_t in the autoregressive part of both regimes (where the notation $X_t^{(i)}$ means that X_t has been generated from regime i).

[ANVss] show that, under the assumption of strictly stationarity and ergodicity of the process X_t, model (2), with $k = 2$, can be alternatively written as

$$X_t = \left[c_0^{(1)} + \sum_{j=0}^{\infty} \psi_j^{(1)} B^j e_t^{(1)} \right] I_{t-d} + \left[c_0^{(2)} + \sum_{j=0}^{\infty} \psi_j^{(2)} B^j e_t^{(2)} \right] (1 - I_{t-d}), \quad (3)$$

with

- $I_{t-d} = \begin{cases} 1 & \text{if } X_{t-d} \geq r_1 \\ 0 & \text{otherwise,} \end{cases}$
- $c_0^{(i)} = \phi_0^{(i)}/(1 - \phi_{p_i}^{(i)}(B)) = \phi_0^{(i)}/(1 - \sum_{j=1}^{p_i} \phi_j^{(i)})., i = 1, 2,$
- $\frac{\theta_{q_i}^{(i)}(B)}{\phi_{p_i}^{(i)}(B)} = \sum_{j=0}^{\infty} \psi_j^{(i)} B^j,$

with $\psi_j^{(i)} = \sum_{s=0}^{j-1} \phi_s^{(i)} \psi_{j-s}^{(i)} - \theta_j^{(i)}$, $\psi_0 = 1$ and $\sum_{j=0}^{\infty} |\psi_j^{(i)}| < \infty$, for $i = 1, 2$.

Starting from model (2), with $k = 2$, and from the alternative form (3), exact multi-step forecasts have been derived. In particular in Section 2, the best multi-step predictor, in terms of minimum mean square error, for this class of models is presented. Different aspects which affect its generation are highlighted further distinguishing among *least squares* and *plug-in* predictors. Taking advantage of these results, the two predictors are properly combined in a scheme based on their variance. In Section 3 an application to two stock markets index returns shows the performance of the proposed predictors when the forecast of the time series level is of interest.

2 The SETARMA Predictors

Given the time series $X_1, \ldots, X_t$ and the lead time h, it is well known that the best predictor of X_{t+h}, obtained from the minimization of the mean square errors forecast, is the conditional expectation $E[X_{t+h}|\Omega_t] = X_t(h)$, where $\Omega_t = \{X_1, \ldots, X_t\}$.

In other words, the least squares predictor is obtained from:

$$\min_{X_t(h)} E[X_{t+h} - X_t(h)]^2,$$

with $X_{t+h} - X_t(h) = e_t(h)$ the forecast error.

When a SETARMA model is involved, the least squares forecast $X_t(h)$ depends on the *threshold variable*, which controls the switching among regimes, and the *threshold delay*, which has relevant implications on the predictor form and on its distribution. More precisely, when $h \le d$, the predictor $X_t(h)$ is derived simply following the results of the linear time series analysis ([BJ76]). Its form is

$$X_t(h) = X_t^{(1)}(h) I_{t+h-d} + X_t^{(2)}(h)(1 - I_{t+h-d}),$$

where $X_t^{(i)}(h)$ is the ARMA predictor in regime i.

Using the SETARMA representation (3), $X_t(h)$ can even be written

$$X_t(h) = \sum_{i=1}^{2} \left[c_0^{(i)} + \sum_{j=h}^{\infty} \psi_j^{(i)} e_{t+h-j}^{(i)} \right] I(X_{t+h-d} \in R_i),$$

where $E(e_{t+h-j}^{(i)} | \Omega_t) = 0$, for $j = 1, ..., h-1$, with $R_1 = [r, +\infty]$ and $R_2 = (-\infty, r)$.

On the contrary, when $h > d$, the estimation of the threshold variable implies some difficulties. In fact, when the lead time h is greater than the threshold delay d, $X_{t+h-d} \notin \Omega_t$ and so I_{t+h-d} becomes a Bernoulli random variable

$$i_{h-d} = \begin{cases} 1 & \text{with} \quad P(X_{t+h-d} \ge r | \Omega_t) \\ 0 & \text{with} \quad P(X_{t+h-d} < r | \Omega_t) \end{cases} \qquad \text{for } h = d+1, d+2, \ldots, \tag{4}$$

where $P(X_{t+h-d} \ge r | \Omega_t) = E[i_{h-d} | \Omega_t] = p_{(h-d)}$.

The least square predictor in this case becomes

$$X_t(h) = X_t^{(2)}(h) + p_{(h-d)} \cdot \left(X_t^{(1)}(h) - X_t^{(2)}(h) \right), \tag{5}$$

which can even be written

$$X_t(h) = c_0^{(2)} + \sum_{j=h}^{\infty} \psi_j^{(2)} e_{t+h-j}^{(1)} + \left[c_0^{(1)} - c_0^{(2)} + \sum_{j=h}^{\infty} \left(\psi_j^{(1)} - \psi_j^{(2)} \right) e_{t+h-j}^{(2)} \right] p_{(h-d)}. \tag{6}$$

The prediction error $e_t(h)$ is

$$e_t(h) = e_t^{(2)}(h) + I_{t+h-d}[e_t^{(1)}(h) - e_t^{(2)}(h)] + [I_{t+h-d} - p_{(h-d)}] \cdot [X_t^{(1)}(h) - X_t^{(2)}(h)], \tag{7}$$

where $e_t^{(i)}(h) = \sum_{j=0}^{h-1} \psi_j^{(i)} e_{t+h-j}^{(i)}$ and $X_t^{(i)}(h) = \sum_{j=h}^{\infty} \psi_j^{(i)} e_{t+h-j}^{(i)}$ are the forecast errorsand the predictions generated from regime i respectively ($i = 1, 2$).

The predictor (6) is unbiased for X_{t+h} as shown in the following proposition.

Proposition 1. *The least square predictor (6) of the strictly stationary and ergodic SETARMA model (2), with $k = 2$ and known coefficients, is an unbiased estimator for X_{t+h}, that is*

$$E[X_t(h)] = X_{t+h} \qquad\qquad h = d+1, \dots \quad . \tag{8}$$

Proof. The unbiasedness of $X_t(h)$ can be equivalently demonstrated showing that the prediction error $e_t(h)$ has $E[e_t(h)] = 0$.
Given $e_t(h)$ in (7), it can even be written as

$$e_t(h) = \sum_{j=0}^{h-1} \psi_j^{(2)} e_{t+h-j}^{(2)} + I_{t+h-d} \left[\sum_{j=0}^{h-1} \psi_j^{(1)} e_{t+h-j}^{(1)} - \sum_{j=0}^{h-1} \psi_j^{(2)} e_{t+h-j}^{(2)} \right] + \\ +(I_{t+h-d} - p_{h-d}) \left(\sum_{j=h}^{\infty} \psi_j^{(1)} e_{t+h-j}^{(1)} - \sum_{j=h}^{\infty} \psi_j^{(2)} e_{t+h-j}^{(2)} \right). \tag{9}$$

Observing that, for $h > d$, I_{t+h-d} is a random variable with distribution (4) and with $E(i_{h-d}) = P(X_{t+h-d} \geq r|\Omega_t) = p_{(h-d)}$, the expected value of (9) is

$$E[e_t(h)] = \sum_{j=0}^{h-1} \psi_j^{(2)} E[e_{t+h-j}^{(2)}] + p_{(h-d)} \left[\left(\sum_{j=0}^{h-1} \psi_j^{(1)} E[e_{t+h-j}^{(1)}] - \right.\right. \\ \left.\left. + \sum_{j=0}^{h-1} \psi_j^{(2)} E[e_{t+h-j}^{(2)}] \right) |I_{t+h-d} \right] = 0,$$

which implies that $E[X_t(h)] = E[X_{t+h} + e_t(h)] = X_{t+h}$. □

The variance of the prediction error $e_t(h)$ is

$$\sigma_e^2(h) = \sigma_{2,e}^2(h) + p[\sigma_{1,e}^2(h) - \sigma_{2,e}^2(h)] + [p + p_{(h-d)}^2 - 2p \cdot p_{(h-d)}] \\ \times[\sigma_{1,X}^2(h) + \sigma_{2,X}^2(h) - 2\sigma_{12,X}(h)]. \tag{10}$$

where $\sigma_{i,X}^2(h) = \sigma_i^2 \sum_{j=h}^{\infty} \left(\psi_j^{(i)}\right)^2$, for $i = 1, 2$, $\sigma_{12,X}(h) = \sigma_1 \sigma_2 \sum_{j=h}^{\infty} \psi_j^{(1)} \psi_j^{(2)}$ is the forecast covariance and p is the unconditional expected value of I_{t+h-d}.

2.1 Plug-in forecasts

When $h > d$ the generation of multi step-ahead predictions can be even accomplished using a different strategy frequently used in empirical framework. If the forecasts are generated treating the values predicted at the previous steps as true values, the conditional set Ω_t grows with the lead time so becoming $\Omega_t(h-d) = \{X_1, \dots, X_t, X_t(1), \dots, X_t(h-d)\}$.

This allows the generation of forecasts conditional to $\Omega_t(h-d)$

$$X_t(h) = E[X_{t+h}|\Omega_t(h-d)] \qquad h = d+1, d+2, \ldots, \tag{11}$$

where the predictions $X_t(1), \ldots, X_t(h-d)$ belong to the conditional set $\Omega_t(h-d)$. The indicator function I_{t+h-d} becomes:

$$i_t(h-d) = [I_{t+h-d}|\Omega_t(h-d)] = \begin{cases} 1 & \text{if } X_t(h-d) \geq r \\ 0 & \text{if } X_t(h-d) < r, \end{cases} \tag{12}$$

with the value of $i_t(h-d)$ related to the forecasts generated at the previous steps.

The predictor (11), called *plug-in* and denoted $X_t^{PI}(h)$ in order to distinguish it from the least square predictor (6) (denoted $X_t^{LS}(h)$), is given as:

$$X_t^{PI}(h) = E[X_{t+h}|\Omega_t(h-d)] = X_t^{(2)}(h) + i_t(h-d)\left(X_t^{(1)}(h) - X_t^{(2)}(h)\right), \tag{13}$$

with forecast error

$$e_t^{PI}(h) = e_t^{(2)}(h) + I_{t+h-d}[e_t^{(1)}(h) - e_t^{(2)}(h)] + [I_{t+h-d} - i_t(h-d)] \cdot [X_t^{(1)}(h) - X_t^{(2)}(h)], \tag{14}$$

where $X_t^{(i)}(h)$ and $e_t^{(i)}(h)$ are defined in the previous section.

It is interesting to note that the predictor (13) is still an unbiased estimator for X_{t+h}.

Proposition 2. *The plug-in predictor $X_t^{PI}(h)$ generated from the strictly stationary and ergodic model (2), with $k = 2$ and known coefficients, is unbiased for X_{t+h}, so*

$$E[X_t^{PI}(h)] = X_{t+h}\,.$$

Proof. The proof follows the same steps of proposition 2. Starting from:

$$e_t^{PI}(h) = X_{t+h} - X_t^{PI}(h), \tag{15}$$

the prediction error $e_t^{PI}(h)$ is

$$\begin{aligned} e_t^{PI}(h) = {} & \sum_{j=0}^{h-1} \psi_j^{(2)} e_{t+h-j}^{(2)} + I_{t+h-d}\left[\sum_{j=0}^{h-1} \psi_j^{(1)} e_{t+h-j}^{(1)} - \sum_{j=0}^{h-1} \psi_j^{(2)} e_{t+h-j}^{(2)}\right] + \\ & +(I_{t+h-d} - i_t(h-d))\left(\sum_{j=h}^{\infty} \psi_j^{(1)} e_{t+h-j}^{(1)} - \sum_{j=h}^{\infty} \psi_j^{(2)} e_{t+h-j}^{(2)}\right), \end{aligned}$$

with unconditional expected value

$$E[e_t^{PI}(h)] = \sum_{j=0}^{h-1} \psi_j^{(2)} E[e_{t+h-j}^{(2)}] + p_{(h-d)} \left[\left(\sum_{j=0}^{h-1} \psi_j^{(1)} E[e_{t+h-j}^{(1)}] + \right.\right.$$
$$\left.\left. - \sum_{j=0}^{h-1} \psi_j^{(2)} E[e_{t+h-j}^{(2)}] \right) |I_{t+h-d} \right] + p_{(h-d)} \left[\left(\sum_{j=h}^{\infty} \psi_j^{(1)} E[e_{t+h-j}^{(1)}] + \right.\right.$$
$$\left.\left. - \sum_{j=h}^{\infty} \psi_j^{(2)} E[e_{t+h-j}^{(2)}] \right) |I_{t+h-d} \right] - i_t(h-d) \left(\sum_{j=h}^{\infty} \psi_j^{(1)} E[e_{t+h-j}^{(1)}] + \right.$$
$$\left. - \sum_{j=h}^{\infty} \psi_j^{(2)} E[e_{t+h-j}^{(2)}] \right) = 0.$$

This implies that the expectation of (15) is

$$E[X_{t+h} - X_t^{PI}(h)] = 0,$$

and so $X_{t+h} = E[X_t^{PI}(h)]$. □

The variance of the prediction error (14) is instead given as

$$\sigma_{PI,e}^2(h) = \sigma_{2,e}^2(h) + p[\sigma_{1,e}^2(h) - \sigma_{2,e}^2(h)] + [p + i_t(h-d) - 2p \cdot i_t(h-d)]$$
$$\times [\sigma_{1,X}^2(h) + \sigma_{2,X}^2(h) - 2\sigma_{12,X}(h)], \quad (16)$$

that, even in this case, is related to the variance of the forecast errors and of the predictors generated from the two regimes.

2.2 Forecasts combination

When two or more predictors are involved, it is advisable to find a criterion which allows their comparison and that gives a measure of their forecast performance. If the selected criterion is the mean square forecast error (MSFE), the comparison of the plug-in and least squares predictors, in the linear context, shows that

$$E[(X_{t+h} - X_t^{PI}(h))^2 | \Omega_t] \geq E[(X_{t+h} - X_t^{LS}(h))^2 | \Omega_t],$$

and, according to the results of proposition 3.4 in [FY03], it is equivalent to state that $\sigma_{PI,e}^2(h) > \sigma_{LS,e}^2(h)$, hence leading to the preference of the least square predictor.

This result is not always true when non-linearity is involved. In particular, given the forecast variances (10) and (16), it can be shown that

$$\frac{\sigma_{LS,e}^2(h)}{\sigma_{PI,e}^2(h)} \geq 1 \qquad \text{if } X_t(h-d) \geq r$$
$$\frac{\sigma_{LS,e}^2(h)}{\sigma_{PI,e}^2(h)} \leq 1 \qquad \text{if } X_t(h-d) < r, \quad (17)$$

Table 1. Forecast accuracy, measured in terms of RMSE=$h^{-1}\sum_{i=1}^{h} e_t^2(i)$, of the Least Squares, Plug-In and Combined predictors generated for X_{t+h}, with $h = 3, \ldots, 10$ (the more accurate forecasts are in bold).

	RMSE($\times 10^{-2}$)		
h	LS	PI	C
3	1.50229	1.53834	**1.47044**
4	**1.35510**	1.41816	1.36402
5	**1.44576**	1.55597	1.50165
6	**1.37401**	1.41586	1.37884
7	1.44292	1.49025	**1.43057**
8	1.40578	1.45710	**1.39949**
9	1.43602	1.45639	**1.41948**
10	1.42430	1.44203	**1.39602**

so highlighting that the ratio between the variances (and the related forecast performances) of the least squares and plug-in forecast errors changes with respect to the regime involved from the threshold variable.

To take advantage of the result of (17), the unbiased predictors (6) and (13) can be combined as follows:

$$\begin{aligned} X_t^C(h) &= i_t(h-d)X_t^{PI}(h) + [1 - i_t(h-d)]X_t^{LS}(h)] \\ &= X_t^{LS}(h) + i_t(h-d)\left[X_t^{PI}(h) - X_t^{LS}(h)\right], \end{aligned} \tag{18}$$

with $i_t(h-d)$ in (12). It is trivial to show that $X_t^C(h)$ is an unbiased forecast for X_{t+h}, when $c_0^{(1)} = c_0^{(2)}$, and the variance of the forecast error $e_t^C(h)$ is such that:

$$\sigma_{C,e}^2(h) = \min\{\sigma_{LS,e}^2(h), \sigma_{PI,e}^2(h)\}. \tag{19}$$

Remark 1. The combination (18) can be seen as selection of the “best” predictor (in term of MSFE) conditional to the augmented set $\Omega_t(h-d)$.

The forecast accuracy of $X_t^C(h)$ has been compared with $X_t^{LS}(h)$ and $X_t^{PI}(h)$ in the following simulated example.

Example 1. Given the SETARMA(2; 1,1; 1,1) model

$$X_t = \begin{cases} 0.8X_{t-1} + e_t^{(1)} - 0.54e_{t-1}^{(1)} & X_{t-2} \geq 0 \\ -0.68X_{t-1} + e_t^{(2)} + 0.36e_{t-1}^{(2)} & X_{t-2} < 0, \end{cases}$$

with $\sigma_1 = 0.5$ and $\sigma_2 = 1$, multi-step ahead forecasts, with $h = 3, \ldots, 10$ have been generated using the three predictors $X_t^{LS}(h)$, $X_t^{PI}(h)$ and $X_t^C(h)$, whose forecast accuracy, evaluated through the Root Mean Square Errors (RMSE), is compared in Table 1.

The results clearly show that, as h increases, the forecast combination outperforms the other two competitors, so affirming the advantage obtained from the combination when h grows.

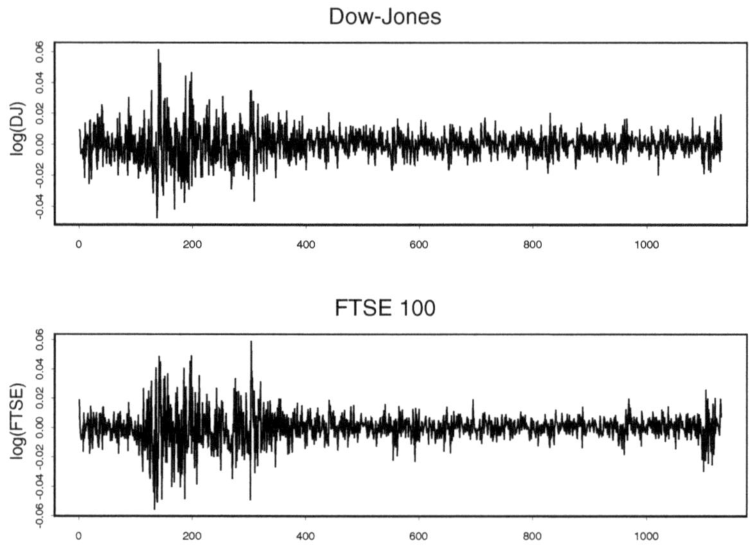

Fig. 1. Dow-Jones and FTSE 100 log-returns from January 1, 2002 to June 30, 2006.

3 Empirical Results and Analysis

In order to evaluate the theoretical results presented and the forecast accuracy of the Least Squares (LS), Plug-in (PI) and Combined (C) predictors in empirical context, they are applied to generate ex-post forecasts from daily Dow-Jones (DJ) and FTSE 100 (FTSE) stock index log-returns from January 1, 2002 to June 30, 2006. The log-returns of the two series are plotted in Fig. 1, where in the first part of both series, an increasing variability can be appreciated.

The SETARMA models fitted to these time series have been selected according to the minimum AIC, after fixing the threshold value at zero (to treat differently positive and negative returns) and leaving out the last 10 observations to generate forecasts. The estimated models are:

$$\text{Dow-Jones:} \qquad X_t = \begin{cases} 0.00055 + e_t^{(1)} - 0.13678 e_{t-1}^{(1)} & X_{t-3} \geq 0 \\ -0.00040 + e_t^{(2)} & X_{t-3} < 0 \end{cases}$$

$$\text{FTSE 100:} \qquad X_t = \begin{cases} 0.00029 + e_t^{(1)} & X_{t-2} \geq 0 \\ -0.00010 + 0.11636 X_{t-3} + e_t^{(2)} & X_{t-2} < 0, \end{cases}$$

from which $h = 10$ step ahead forecasts have been generated using the three predictors presented.

Their accuracy has been compared, in terms of RMSE, in Table 2 where in both time series, the combination outperforms the remaining two forecasts.

Table 2. RMSE($\times 10^{-2}$) of the forecasts generated from DJ and FTSE time series with $h = 10$.

	LS	PI	C
Dow-Jones	0.86475	0.86189	0.86021
FTSE 100	0.74898	0.74271	0.74223

This confirms the results obtained in the simulated example 1, but with the no negligible detail that now the model parameters are estimated, and so estimation errors affect the forecast accuracy.

Starting from the theoretical results presented, more complex procedures of forecasts combination can be defined. In fact, as remarked in Section 2.2, the proposed combination can be intended as a "predictor selection". This can be refined for example using combination schemes based on proper weighted means of predictors.

References

[ANVss] A. Amendola, M. Niglio, and C. Vitale. The autocorrelation functions in setarma models. In E. J. Kontoghiorghes and C. Gatu, editors, *Optimization, Econometric and Financial Analysis*. Springer-Verlag (2007)

[BJ76] G. E. P. Box and G. M. Jenkins. *Time series analysis, forecasting and control.* Holden-Day, San Francisco (1976)

[FY03] J. Fan and Q. Yao. *Nonlinear time series. Nonparametric and parametric methods.* Springer-Verlag (2003)

[Ton78] H. Tong. On a threshold model. In C.H. Chen, editor, *Pattern recognition and signal processing*. Sijthoff and Noordhoff, Amsterdam (1978)

[Ton83] H. Tong. *Threshold models in nonlinear time series analysis*. Springer-Verlag (1983)

Estimating Portfolio Conditional Returns Distribution Through Style Analysis Models*

Laura Attardi and Domenico Vistocco

Summary. Style analysis models aim to decompose the performance of a financial portfolio with respect to a set known indexes. Quantile regression offers a different point of view on the style analysis problem as it allows the extraction of information at different parts of the portfolio returns distribution. Moreover, the quantile regression results are useful in order to estimate the portfolio conditional returns distribution.

Key words: Style analysis; Quantile regression, Portfolio conditional returns distribution.

1 Introduction

The aim of this paper is to investigate how to exploit the information provided by quantile regression models (for different values of conditional quantiles) in order to enrich the classical results carried by the least squares style analysis model. The quantile regression approach allows an indepth investigation of the impact of asset class exposure on portfolio conditional returns distribution. Moreover, portfolio conditional returns distribution can be estimated by exploiting quantile regression estimates. In this way, inferences on portfolio returns can be conducted in a semi-parametric framework, without the need to assume the usual conditions that are used in common practice.

The paper is organized as follows. In Section 2 we briefly introduce the style analysis problem. An illustrative portfolio is described in Section 2.1: it is used to show the least squares approach and the quantile regression approach (Section 2.2) to style analysis. The quantile regression results are then used to estimate portfolio conditional returns distribution. Some concluding remarks follow in Section 3.

* All computation and graphics were done in the R language (www.r-project.org) [R07] using the basic packages and the additional mgcv [WOO06] and quantreg packages [KOE07].

Cira Perna, Marilena Sibillo (Eds.): Mathematical and Statistical Methods in Insurance and Finance

2 Style Analysis

The style analysis model decomposes the portfolio performance with respect to a set of known indexes (constituents). The constituents are typically selected to represent the different asset classes in which the portfolio invests. The model estimates the quotas of such indexes in the portfolio, with the aim of separating out their share to return.

The classical model is based on a least squares constrained regression model [SHA92][SHA98]. Let us denote by $\mathbf{r}^{port}$ the vector of portfolio returns along time and $\mathbf{r}^{const}$ the matrix containing the returns along time of the ith portfolio constituent on the ith column ($i = 1, \ldots, n$). Data are observed on T subsequent time periods. The style analysis model regresses portfolio returns on the returns of the n constituents:

$$\mathbf{r}^{port} = \mathbf{r}^{const}\mathbf{w}^{const} + \mathbf{e}$$

subject to: $\mathbf{w}^{const} \geq 0,\ \mathbf{1}^T\mathbf{w}^{const} = 1$.

The vector $\mathbf{e}$ denotes the tracking error of the portfolio. The two constraints allow the coefficients to be exhaustive and non-negative, allowing them to be interpreted in terms of compositional data.

2.1 An illustrative portfolio

In order to illustrate how to exploit the information provided by least squares (LS) and quantile regression (QR) models, an application on a bond portfolio follows. The portfolio has been obtained as a combination of G7 Merril Lynch indexes: they consist of indexes tracking the performance of the oustanding public local currency debt of G7 sovereign issuers. In particular, they refer to Canadian, French, German, Italian, Japanese, UK and US sovereign bonds issued in their respective domestic markets [MER06]. The time series ($T = 209$ observations) was quoted in Euro. Weekly data were used from 7 April 2001 to 23 February 2005, using the Wednesday observation as common practice in financial data analysis.

Table 1. Summary statistics for the portfolio.

	portfolio	constituents weights						
	returns	CAN	FRA	GER	ITA	JAP	UK	USA
min	-0.0258	0.0788	0.0834	0.0316	0.1111	0.0524	0.0159	0.1333
Q1	-0.0061	0.1267	0.1032	0.0412	0.1425	0.0882	0.0209	0.1957
median	-0.0007	0.1578	0.1376	0.0537	0.1652	0.1149	0.0287	0.3310
mean	-0.0004	0.1565	0.1329	0.0755	0.1717	0.1238	0.0478	0.2919
Q3	0.0046	0.1961	0.1563	0.1147	0.2079	0.1478	0.0813	0.3543
max	0.0374	0.2644	0.1829	0.2004	0.2444	0.3120	0.1429	0.4251
σ	0.0100	0.0483	0.0282	0.0446	0.0346	0.0507	0.0341	0.0865

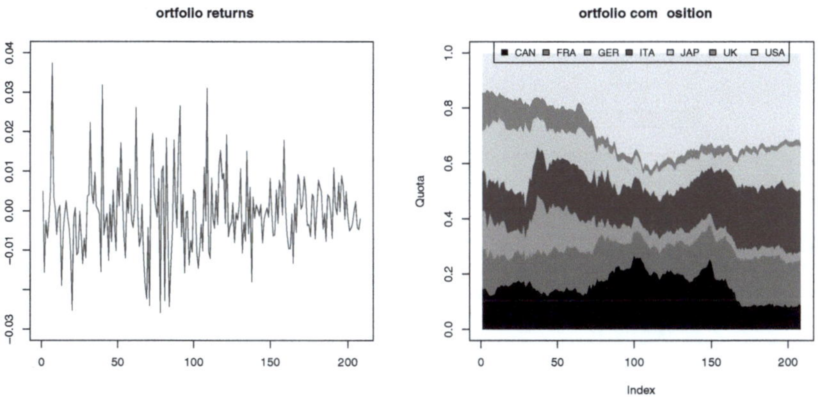

a **b**

Fig. 1. A simulated bond portfolio: **a** returns along time, and **b** constituent quotas along time.

Portfolio constituent weights along time have been obtained using a classic optimization procedure [ELT95], so mimicking a typical active portfolio. Summary statistics for the obtained portfolio are contained in Table 1. The first column table refers to portfolio returns while the others contain summary indicators for constituent weights. Fig. 1a shows the portfolio returns while the constituent quotas in composing the portfolio are depicted in Fig. 1b.

2.2 The least squares model and the quantile regression model

The use of LS model focuses on the conditional expectation of portfolio returns distribution, as the LS model can be formulated as follows:

$$E(\mathbf{r}^{port} \mid \mathbf{r}^{const}) = \mathbf{r}^{const}\mathbf{w}^{const}$$

with constraints: $\mathbf{w}^{const} \geq 0,\ \mathbf{1}^T\mathbf{w}^{const} = 1$.

Estimated compositions are then interpretable in terms of sensitivity of portfolio expected returns to constituent returns. In the classical regression context, the $w_t^{const_i}$ coefficient represents the impact of a change in the returns of the ith constituent on the portfolio expected returns, holding the values of the other constituent returns constant. Using the LS model, portfolio style is then determined estimating the style exposure influence on expected returns. Fig. 2 depicts least squares results for the bond portfolio. Fig. 2a shows the real portfolio composition at time T compared with the estimated composition. Although the estimated LS coefficients are in common practice interpreted as composition at the last time, it is worthwhile stressing that they reflect the average effect along the period, according to the regression framework. This is evident from Fig. 2b where estimated LS coefficients are compared to average portfolio composition.

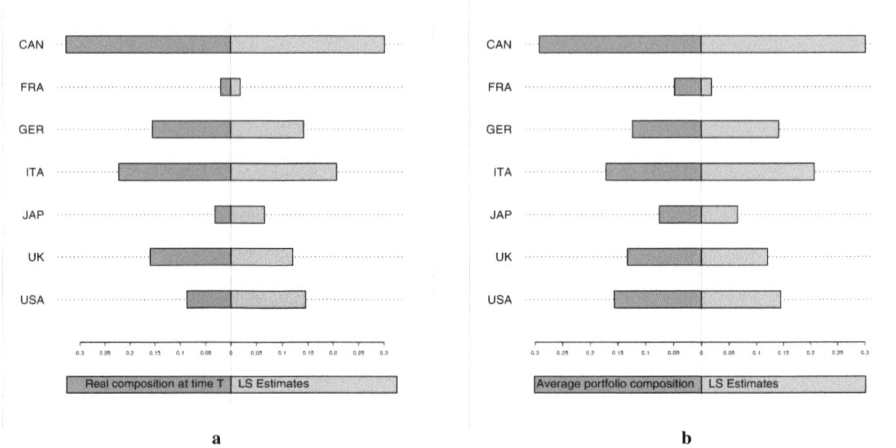

Fig. 2. Style analysis coefficients compared: **a** to the real portfolio composition at time T, and **b** to the average portfolio composition.

Exploiting quantile regression [KOE05], a more detailed comparison of financial portfolios is obtainable as quantile regression coefficients are interpretable in terms of sensitivity of portfolio conditional quantiles returns to constituent returns [BAS01]. The quantile regression (QR) model for a given conditional quantile θ follows:

$$Q_\theta(\mathbf{r}^{port} \mid \mathbf{r}^{const}) = \mathbf{r}^{const}\mathbf{w}^{const}(\theta)$$

with constraints: $\mathbf{w}^{const}(\theta) \geq 0$, $\mathbf{1}^T\mathbf{w}^{const}(\theta) = 1, \forall\theta$, where θ $(0 < \theta < 1)$ denotes the particular quantile of interest [ATT06].

In a similar way, as for the LS model, the $w_t^{const_i}(\theta)$ coefficient of the QR model can be interpreted as the rate of change of the θth conditional quantile of the portfolio returns distribution for a one unit change in the ith constituent returns holding the values of $X_{j,j\neq i}$ constant. The use of QR then offers a more complete view of relationships among portfolio returns and constituent returns.

Table 2 collects results of LS model and of QR models for a set of selected quantiles ($\theta = \{0.05, 0.10, 0.25, 0.50, 0.75, 0.90, 0.95, 0.99\}$): the seven portfolio constituents are in rows while the columns refer to the different models, the last column being the difference between maximum and minimum value for each row. Besides the LS coefficients, QR coefficients provide information on the different level of turnover in the portfolio constituents along time.

The obtained results can be graphically inspected. The graphical representations allow a better appreciation of the different impact of the sectors composing the portfolio on the total return. Fig. 3 displays, with respect to the seven sectors, the QR estimates for the analyzed portfolio. The different conditional quantiles are represented on the X axis while the coefficient values are on the Y axis, the solid line with filled dots representing the point estimates for 99 distinct conditional quantiles ($\theta = \{0.01, 0.02, \ldots, 0.98, 0.99\}$). From the figure, it is evident that there is a differ-

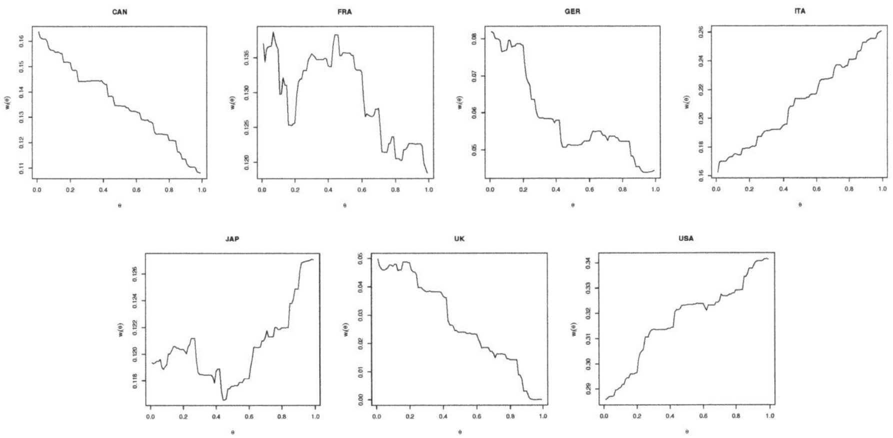

Fig. 3. Quantiles style analysis coefficients for the seven different sectors.

ent pattern for CAN, GER and UK sectors from one side and for ITA and USA on the other side. The first show a decreasing effect (lower quantiles correspond to upper estimates and vice versa) while the others show an increasing effect (lower quantiles correspond to lower estimates and upper quantiles correspond to upper estimates). Finally FRA and JAP sectors show a different behaviour with respect to the net effect of the correspondent returns on portfolio conditional quantile returns. Moreover, exploiting the conditional quantile estimates shown in Fig. 3, it is straightforward to estimate the portfolio conditional returns distribution as follows:

$$\hat{\mathbf{r}}_{\theta}^{port} = \mathbf{r}^{const} \hat{\mathbf{w}}^{const}(\theta), \qquad \text{for } 0 < \theta < 1.$$

Table 2. LS and QR estimates for the analyzed portfolio.

portfolio	LS	Q(0.05)	Q(0.10)	Q(0.25)	Q(0.50)	Q(0.75)	Q(0.90)	Q(0.95)	max - min
CAN	14.50%	16.09%	15.63%	14.84%	13.44%	12.34%	11.35%	11.04%	5.05%
FRA	12.02%	13.66%	13.62%	13.20%	13.57%	12.14%	12.27%	12.26%	1.64%
GER	6.57%	7.99%	7.69%	6.75%	5.12%	5.37%	4.54%	4.39%	3.60%
ITA	20.69%	17.01%	17.34%	18.11%	21.37%	23.71%	25.27%	25.54%	8.53%
JAP	14.16%	11.95%	11.92%	12.07%	11.76%	12.13%	12.49%	12.69%	2.40%
UK	1.89%	4.60%	4.73%	4.45%	2.40%	1.63%	0.30%	0.00%	4.73%
USA	30.16%	28.71%	29.07%	30.57%	32.34%	32.69%	33.78%	34.08%	5.37%
	100%	100%	100%	100%	100%	100%	100%	100%	

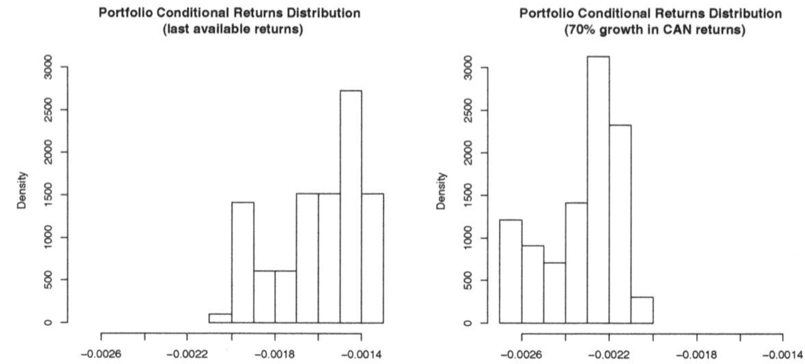

a **b**

Fig. 4. Portfolio conditional returns distribution conditioning using as covariate values *(***a***)* the last constituent returns, and *(***b***)* with a 70% growth in CAN returns.

The obtained distribution is strictly dependent on the values used for the co-variates. It is then possible to use different potential scenarios in order to evaluate the effect on the portfolio conditional returns distribution, carrying out a what-if analysis. Fig. 4a depicts the estimated portfolio conditional returns distribution obtained using the constituent returns at time T as co-variate values while Fig. 4b was obtained using the last available constituent returns but supposing a 70% growth in CAN returns. In this way it is possible to evaluate how the entire shape of the conditional density of the response changes with different values of the conditioning co-variates, without confining oneself to the classical regression assumption that the co-variates affect only the location of the response distribution, but not its scale or shape.

3 Concluding Remarks

Style analysis models allow information to be obtained on the impact of exposure choices on portfolio returns. The classical least squares model estimates the style exposure influence on portfolio expected returns. The quantile regression approach allows the extraction of information at places other than the expected value, providing information on the influence of exposure choices on the entire conditional returns distribution. Therefore, quantile regression can be used as a complement to standard analysis, allowing discrimination among portfolios that would be otherwise judged equivalent using only conditional expectation. Furthermore, through quantile regression estimates it is possible to obtain information on the portfolio conditional returns distribution. It can be estimated starting from a hypothesis on constituent returns (e.g. last available returns) or by supposing different scenarios in order to compare the resulting distributions. Further investigation to make the best use of quantile regression potential for style analysis should concern the simulation of a more numerous set of portfolios. Moreover, the method should be tested on portfolio investing in different asset classes.

References

[ATT06] Attardi, L., Vistocco D.: Comparing financial portfolios style through quantile regression, Statistica Applicata, 18(2) (2006)

[BAS01] Basset, G. W., Chen, H. L.: Portfolio Style: Return-Based Attribution Using Quantile Regression. In: Fitzenberger B., Koenker R. and Machado J. A. F. (eds) Economic Applications of Quantile Regression (Studies in Empirical Economics). Physica-Verlag, 293–305 (2001)

[ELT95] Elton, E. J., Gruber, M.J.: Modern Portfolio Theory and Investment Analysis. John Wiley & Sons, Chichester (1995)

[KOE05] Koenker, R.: Quantile Regression. Econometric Soc. Monographs (2005)

[KOE07] Koenker, R.: quantreg: Quantile Regression. R package version 4.06. http://www.r-project.org (2007)

[R07] R Development Core Team: R: A language and environment for statistical computing. R Foundation for Statistical Computing, Vienna, Austria. ISBN 3-900051-07-0, URL http://www.R-project.org (2007)

[MER06] Merril Lynch: MLIndex System. http://www.mlindex.ml.com (2006)

[SHA92] Sharpe, W.: Asset Allocation: Management Styles and Performance Measurement. The Journal of Portfolio Management (1992)

[SHA98] Sharpe, W.: Determining a Fund's Effective Asset Mix. Investment Management Review, vol. 2, no. 6, December:59–69 (1998)

[WOO06] Wood, S. N.: Generalized Additive Models: An Introduction with R. Chapman and Hall/CRC (2006)

A Full Monte Carlo Approach to the Valuation of the Surrender Option Embedded in Life Insurance Contracts*

Anna Rita Bacinello

Summary. In this paper we extend the Least Squares Monte Carlo approach proposed by Longstaff and Schwartz for the valuation of American-style contingent-claims to the case of life insurance contracts. These contracts, in fact, often embed an American-style option, called surrender option, that entitles its owner to terminate early the contract and receive a cash amount, called surrender value. The additional complication arising in life insurance policies with respect to purely financial American contracts is that there is not a fixed date within which the option can be exercised, since its "maturity" is driven by mortality factors. This complication has been handled by very few papers, often at the cost of excessively simplified valuation frameworks. The aim of this contribution, that is not a specific valuation model but a methodological approach, is to allow a full exploitation of the flexibility inborn in Monte Carlo and quasi-Monte Carlo methods in order to deal with more realistic valuation frameworks.

Key words: Surrender option; Least Squares Monte Carlo approach.

1 Introduction

The surrender option embedded in several types of life insurance contracts gives the policyholder the right to terminate early the contract, before its natural termination (that is typically *death* or *maturity*), and to receive a cash amount, called *surrender value*. It is a non-standard *knock-out* American put option written on the residual contract, with exercise price given by the surrender value. The knock-out feature is implied by the fact that this option can be exercised only if the insured is still alive, hence it expires in case of death ("knock-out event"). The additional complication arising in life insurance policies with respect to standard American options is that there is not a fixed date within which the option can be exercised, since its "maturity" is driven by mortality factors. Moreover, the value of the residual contract depends both on mortality and on financial uncertainty and, even if pooling arguments can be

* Financial support from Cofinanziamento MIUR on 'Valutazioni nelle assicurazioni vita e nel settore previdenziale: metodologie attuariali, economiche e finanziarie" is gratefully acknowledged.

Cira Perna, Marilena Sibillo (Eds.): Mathematical and Statistical Methods in Insurance and Finance

applied in order to hedge the mortality risk, it is not possible, in the valuation, to keep these two sources of uncertainty separate because there is a continuous interaction between them.

The literature concerning the valuation of the surrender option in a contingent-claims framework is not very abundant, and most of the papers on this subject deal with purely financial contracts, without mortality risk, applying to them the results on American options. There are very few exceptions that deal with actual life insurance contracts, characterized by both financial and mortality risk. However, the complexity of the problem involved in this case often forces the assumption of excessively simplified valuation frameworks (deterministic interest rates, deterministic volatility for reference portfolios, deterministic mortality trends).

The valuation approaches to tackle the problem can be essentially classified in three categories:

1. Binomial/multinomial trees (see [Bac03a, Bac03b, Bac05, Van03a, Van03b]).
2. Partial Differential Equations with free boundaries (see [SX05]).
3. Monte Carlo simulation (see [AC03, BDF06]).

In particular, the papers [AC03] and [BDF06] combine the Least Squares Monte Carlo approach (LSM henceforth) proposed by [LS01] for the valuation of *purely-financial* American-style contingent-claims with the approach proposed by [Bac03a] and [Bac03b] to manage the mortality risk in the valuation of the surrender option. They follow the LSM approach only to handle the financial uncertainty (stochastic interest rates, stochastic evolution of reference portfolios), but resort to "analytic" tools for the mortality one. Then the two sources of uncertainty are not treated in the same way and the stochastic time of death does not enter either the simulation process or the definition of the stochastic discounted cash-flow of the contract.

This paper extends instead the LSM approach to the case of life insurance contracts in a very natural way, according to which the mortality uncertainty is treated exactly as the financial one and is part of the whole LSM mechanism. The aim of this contribution, that is not a specific valuation model but a methodological approach, is to allow a full exploitation of the flexibility inborn in Monte Carlo and quasi-Monte Carlo methods, in order to deal with very realistic valuation frameworks including for example stochastic interest rates, stochastic volatility, jumps, stochastic mortality trends, etc.

The paper is structured such that in Section 2 we present our notation and assumptions, in Section 3 we describe the valuation approach and in Section 4 we discuss the numerical accuracy of its results. Finally, Section 5 concludes the paper.

2 Notation and Assumptions

Consider an endowment policy issued at time t_0 and maturing at time t_N or, alternatively, a whole-life assurance policy issued at time t_0. In both cases, assume that the policy is paid by a single premium at issuance, denoted by U. Assume moreover that the endowment policy can be surrendered at the discrete dates $t_1, t_2, \dots, t_{N-1}$ before maturity, if the insured is still alive. Similarly, the whole-life assurance policy can be surrendered at the discrete dates t_n, $n = 1, 2, \ldots$, belonging to a given set and before death of the insured. In the case of death between times t_{n-1} and t_n, $n = 1, 2, \ldots$, the benefit is assumed to be paid at the end of the interval, that is at time t_n. Of course, if we consider an endowment policy and the insured is still alive at maturity t_N, the benefit is paid at maturity.

Observe that, to avoid adverse selection, the surrender option is usually offered only when a life insurance contract provides benefits with certainty (even if their amount and/or the time at which they are due are uncertain). That is why we limit ourselves to consider these two types of contracts, where the benefit is due with certainty, soon or later.

Assume that the contract under scrutiny is a unit-linked or a participating policy characterized, as usual, by a rather high level of financial risk. The benefit and the surrender value can then be linked to the value or to the performance of a reference portfolio, with possible minimum guarantees, or upper bounds. Here it is not important to specify their structure. We only denote by F_{t_n}, $n = 0, 1, 2, \ldots$, the value of the reference portfolio at time t_n and by R_{t_n}, $n = 1, 2, \ldots$, the surrender value paid in the case of surrender at time t_n. The "payoff" of the contract is then given by the benefit, at death or maturity, or the surrender value, in the case of surrender.

3 The Valuation Approach

In this section we describe our valuation approach. This approach concerns the whole contract, including the surrender option. Its output is the *fair value* of the contract at time 0, that is also the fair single premium to require for it. Note that with a drastic simplification, that we briefly describe at the end of the section, the same procedure provides the single premium for the corresponding European version of the contract, that is without the surrender option. If one is interested in valuating separately this option, the simplest way to do it is to compute the difference between the time 0 value of the whole contract and that of its European version.

The valuation procedure can be schematized in the following steps.

Step 1. Generate a certain number, say H, of (independent) simulated values of the remaining lifetime of the insured. Then, with reference to the h-th iteration ($h = 1, 2, \ldots, H$):

- let $T^{(h)}$ denote the simulated time at which the benefit would be due if the surrender option were never exercised;
- produce a simulated path of the stochastic term-structure of spot interest rates for any possible maturity up to time $T^{(h)}$;
- produce a simulated path of the reference portfolio at times $t_1, t_2, \ldots, T^{(h)}$.

Remarks: All simulations must be performed under a *risk-neutral* measure Q, taking into account all possible dependencies. For instance, a path for the volatility of zero-coupon bond prices and of the reference portfolio has also to be simulated if these volatilities are assumed to be stochastic. Moreover, if we assume a diffusion process to describe the value of the reference portfolio, its drift under the *risk-neutral* measure is given by the instantaneous spot rate, and so on. Similarly, if future mortality trends are assumed to be stochastic, the simulation of the remaining lifetime of the insured requires the prior simulation of all variables on which it depends, such as the instantaneous force of mortality.

Step 2. Let $t_{max} = \max\left\{T^{(h)} : h = 1, 2, \ldots, H\right\}$. Then, with reference to all iterations (h) such that $T^{(h)} = t_{max}$, determine the "final" payoff of the contract, given by the simulated benefit at death or maturity, and denote it by $P_{t_{max}}^{(h)}$.

Remark: If we are dealing with an endowment policy, it is very likely that $t_{max} = t_N$.

Step 3. Let $n = max - 1, max - 2, \ldots, 1$.

- With reference to all iterations (h) such that $T^{(h)} = t_n$, let $P_{t_n}^{(h)}$ be given by the corresponding simulated benefit.
- With reference to all iterations (h) such that $T^{(h)} > t_n$, use the Least Squares method to estimate

$$E_{t_n}^{Q}\left[\sum_{j:t_n<t_j\leq T} P_{t_j} v\left(t_n, t_j\right)\right],$$

  where $E_{t_n}^{Q}$ denotes expectation, under the (chosen) risk-neutral measure Q, conditioned to the information available up to time t_n and to the event that the contract is still in force at this time (i.e., insured still alive and contract not surrendered yet). T denotes the stochastic date at which the benefit would be due, P_{t_j} denotes the stochastic future payoff of the contract at time t_j and $v\left(t_n, t_j\right)$ denotes the stochastic discount factor from t_j to t_n at the riskless rate.

  Then, denoting by $f_{t_n}^{(h)}$ the estimated conditional expectation, compare it with the corresponding simulated surrender value $R_{t_n}^{(h)}$:
  if $R_{t_n}^{(h)} \leq f_{t_n}^{(h)}$, let $P_{t_n}^{(h)} = 0$ and do not change the future payoffs $P_{t_j}^{(h)}$, $j > n$;
  if $R_{t_n}^{(h)} > f_{t_n}^{(h)}$, let $P_{t_n}^{(h)} = R_{t_n}^{(h)}$ and $P_{t_j}^{(h)} = 0$ for any $j > n$.

Remarks: This step requires a choice of the basis functions whose linear combination defines the regression function, as well as all relevant variables (e.g., the current value of the reference portfolio, the current spot rate, the current force of mortality, ... , and their past values if the model assumed is not Markovian). Note that, inside the Q-expectation, we have the sum of all future payoffs. There is actually one and only one non-zero future payoff, because at any time t_j P_{t_j} can be expressed as the benefit, at death or maturity, or as the surrender value, multiplied for an indicator function that is equal to 1 only once, when the benefit or the surrender value is paid. The data used in the regression are given by the simulated discounted payoffs in all iterations (k) such that $T^{(k)} > t_n$, i.e. by

$$\sum_{j:t_n<t_j\leq T^{(k)}} P_{t_j}^{(k)} v^{(k)}\left(t_n, t_j\right),$$

where $v^{(k)}$ denotes the simulated discount factor at the riskless rate in the k-th iteration. As happens for the corresponding random variable, $P_{t_j}^{(k)} \geq 0$ for any $j > n$ and there exists a unique j such that $P_{t_j}^{(k)} > 0$. Finally, we observe that in [LS01], with reference to a Bermudan-style put option, the authors recommend limiting the application of the regression step only to the simulated paths in which the option is in-the-money. Here, instead, we have to consider all paths because, first of all, we are valuing the whole contract and not simply the surrender option and, secondly, it is not possible to establish if this option is in-the-money since its underlying variable, the value of the residual contract, is not observable but could only be estimated by means of the same procedure under execution.

Step 4. The fair value of the whole contract at time 0, and hence the fair single premium, is given by

$$U = \frac{1}{H} \sum_{h=1}^{H} \sum_{j:\, t_0 < t_j \leq T^{(h)}} P_{t_j}^{(h)} v^{(h)} \left(t_0, t_j\right).$$

Remark: The premium for the corresponding European version of the contract can be simply computed as the average, over all iterations (h), of the simulated benefit paid at time $T^{(h)}$ discounted up to time t_0.

4 Tests of Accuracy

The numerical accuracy of the method here proposed has been verified with reference to the valuation framework assumed in [Bac05]. This is a very simple framework, in which there is a single state-variable given by the value of the reference portfolio, that follows the binomial model by [CRR79]. However, this is the only one in which we have exact analytic results to compare with those produced by the Monte Carlo approach: that is why we have chosen it to check the accuracy of this approach.

We recall that the contract analyzed in [Bac05] is an equity-linked endowment policy. In particular, given a constant length Δ for each time interval $\left[t_{n-1}, t_n\right]$, let $t_0 = 0$ and $t_n = n\Delta$, $n = 1, 2, \ldots, N$. Then, conditionally to the current level of the reference portfolio at time t_{n-1}, given by $F_{t_{n-1}}$, its level at time t_n can take only two possible values, respectively given by $F_{t_{n-1}}u$ and $F_{t_{n-1}}d$, with (risk-neutral) probabilities $q = (\exp(r\Delta) - d) / (u - d)$ and $1 - q = (u - \exp(r\Delta)) / (u - d)$, where r denotes the instantaneous riskless rate on an annual basis (deterministic and constant), $u = \exp(\sigma\sqrt{\Delta})$, $d = 1/u$, and σ represents the volatility of the reference portfolio, once again on an annual basis.

We have done a very large amount of numerical experiments, following different approaches to produce the simulated path of the reference portfolio, with different sets of parameters and by using both pseudo-random numbers and multidimensional low-discrepancy sequences. As basis functions we have employed either powers or Laguerre polynomials.

Observe that in each iteration (h), once the stochastic date T at which the benefit is due has been simulated, the simulated path of the reference portfolio can be generated

- *Forwards*, from time t_1 to $T^{(h)}$, by using the conditional distribution above recalled.
- *Backwards*, by simulating first the value of the reference portfolio at time $T^{(h)}$ from a binomial distribution, and after its values between times $T^{(h)}$ and t_1 from the corresponding conditional distributions. To this end we recall that, given $T^{(h)} = j\Delta$, $j = 1, 2, \dots, N$, the possible values of $F_{T^{(h)}}$ are $F_0 u^i d^{j-i}$, $i = 0, 1, \dots, j$, with (binomial) risk-neutral probability $\binom{j}{i} q^i (1-q)^{j-i}$. Given instead the level at time t_n of the reference portfolio, $F_{n\Delta} = F_0 u^i d^{n-i}$, $n = 2, 3, \dots, N$ and $i = 0, 1, \dots, n$, its level at time t_{n-1} can take only two possible values, that are $F_{n\Delta}/u$ and $F_{n\Delta}/d$, with probabilities i/n and $1 - i/n$. Note that if $i = 0$ or $i = n$, it actually takes only one value, given respectively by $F_{n\Delta}/d$ and $F_{n\Delta}/u$, and hence in these cases no simulations are required.

In particular, the accuracy of the results obtained by following these two different approaches has turned out to be

- The same, when pseudo-random numbers are employed.
- Better backwards than forwards, when using low-discrepancy sequences (and for some sequences, e.g. the Halton one, very much better).

However, since Monte Carlo methods have a very low convergence speed compared with quasi-Monte Carlo, and hence require a larger number of iterations in order to achieve the desired precision, it is better to use the backward approach also here. In fact, this approach does not require a record of all the entire simulated paths of the state-variable, but only of its last simulated values, so that the spared allocated memory can be used in order to increase the number of iterations.

In Table 1 we show the results of some numerical examples. In all of them we have fixed $t_N = 10$ years, $\Delta = 1/12$, so that the length of each time interval is one month and the relevant dates are $t_n = n/12$, $n = 0, 1, 2, \dots, 120$. The insured is assumed to be 40 years old at time 0, and the mortality probabilities used to simulate his or her residual lifetime are extracted from the life table of the Italian Statistics for Female Mortality in 2001, with values corresponding to non-integer ages computed by linear interpolation. The initial value of the reference portfolio, F_0, is set equal to 1. Both the benefit paid at time t (death or maturity) and the surrender value R_t are assumed to be given by $\max\{F_t, F_0 \exp(gt)\}$, so that g represents an instantaneous minimum interest rate guaranteed. We have chosen a constant and a certain number of Laguerre polynomials as basis functions. This number has been fixed in such a way as to maximize the accuracy of the results, within a maximum of 13. The number of iterations H, instead, has been fixed in such a way that the premium for the corresponding European version of the contract computed by means of the simplified valuation procedure described at the end of the previous section coincides with the exact premium analytically computed (at least until the 4th decimal digit).

Table 1. Numerical examples.

r	g	σ	number of iterations H	number of Lag. pol.	estimated premium U	exact premium	error
0.03	0.00	0.15	100,000	9	1.0976	1.0976	0.0000
0.03	0.00	0.25	100,000	10	1.1971	1.1969	0.0002
0.03	0.00	0.35	1,000,000	13	1.2945	1.2956	−0.0011
0.03	0.02	0.15	100,000	11	1.1465	1.1464	0.0001
0.03	0.02	0.25	100,000	10	1.2604	1.2603	0.0001
0.03	0.02	0.35	1,000,000	13	1.3652	1.3674	−0.0022
0.05	0.00	0.15	100,000	9	1.0698	1.0700	−0.0002
0.05	0.00	0.25	100,000	9	1.1555	1.1555	0.0000
0.05	0.00	0.35	1,000,000	13	1.2444	1.2449	−0.0005
0.05	0.02	0.15	100,000	9	1.0975	1.0977	−0.0002
0.05	0.02	0.25	100,000	8	1.1964	1.1972	−0.0008
0.05	0.02	0.35	1,000,000	13	1.2938	1.2955	−0.0017
0.05	0.04	0.15	100,000	7	1.1458	1.1460	−0.0002
0.05	0.04	0.25	100,000	7	1.2599	1.2598	0.0001
0.05	0.04	0.35	1,000,000	13	1.3635	1.3673	−0.0038

In this way the error inborn in the Monte Carlo method is quite negligible and the residual error is mainly due to the regression. Finally, these results are obtained by using the multidimensional Weyl low-discrepancy sequence, that behaves very well despite being the simplest one.

Finally a few comments about our findings. First of all note that, as expected, the number of iterations required to achieve the desired precision for the European version of the contract increases with the volatility of the reference portfolio. In particular, to produce the numerical results of Table 1 we have carried out 100, 000 iterations when the volatility parameter σ equals 0.15 or 0.25, and 1, 000, 000 iterations when $\sigma = 0.35$. The average absolute error is equal to 18.6 basis points (bp) when the volatility is high, to 2.4 bp when it is medium and only to 1.4 bp when it is low. Moreover, the number of Laguerre polynomials required to optimize the regression procedure is equal to 9 (on average) when the volatility parameter σ equals 0.15 or 0.25, and always to the maximum number here fixed, 13, when $\sigma = 0.35$. This indicates that, in case of high volatility, it would be better to further increase the number of iterations and/or the number of basis functions. To summarize, we can conclude that the results obtained are in general very good, and therefore the methodological approach here proposed seems to be suitable for application to more realistic valuation frameworks, with stochastic interest rates, stochastic volatility, jumps, stochastic mortality trends, etc.

5 Summary and Conclusions

In this paper we have proposed a method to extend the Longstaff-Schwartz Least Squares Monte Carlo approach to the case of life insurance contracts embedding a surrender option. We have then applied it to a very simple framework, in which there are exact analytic solutions, in order to verify its accuracy. The accuracy tests indicate that this method performs very well, especially for low or medium-volatile reference portfolios, and hence the next step is to apply it to more sophisticated frameworks, even if there is no way to compare the results obtained with exact ones.

References

[AC03] Andreatta, G., Corradin, S.: Valuing the Surrender Option Embedded in a Portfolio of Italian Life Guaranteed Participating Policies: A Least Squares Monte Carlo Approach. Working Paper (2003)

[Bac03a] Bacinello, A. R.: Fair Valuation of a Guaranteed Life Insurance Participating Contract Embedding a Surrender Option. *The Journal of Risk and Insurance*, **70**, 461–487 (2003)

[Bac03b] Bacinello, A. R.: Pricing Guaranteed Life Insurance Participating Policies with Annual Premiums and Surrender Option. *North American Actuarial Journal*, **7**, 1–17 (2003)

[Bac05] Bacinello, A. R.: Endogenous Model of Surrender Conditions in Equity-Linked Life Insurance. *Insurance: Mathematics and Economics*, **37**, 270–296 (2005)

[BDF06] Baione, F., De Angelis, P., Fortunati, A.: On a Fair Value Model for Participating Life Insurance Policies. *Investment Management and Financial Innovations*, **3**, 105–115 (2006)

[CRR79] Cox, J. C., Ross, S. A., Rubinstein, M.: Option Pricing: A Simplified Approach. *Journal of Financial Economics*, **7**, 229–263 (1979)

[LS01] Longstaff, F. A., Schwartz, E. S.: Valuing American Options by Simulation: A Simple Least-Squares Approach. *The Review of Financial Studies*, **14**, 113–147 (2001)

[SX05] Shen, W., Xu, H.: The Valuation of Unit-Linked Policies with or without Surrender Options. *Insurance: Mathematics and Economics*, **36**, 79–92 (2005)

[Van03a] Vannucci, E.: The Valuation of Unit Linked Policies with Minimal Return Guarantees under Symmetric and Asymmetric Information Hypotheses. Report n. 237, Dipartimento di Statistica e Matematica Applicata all'Economia, Università degli Studi di Pisa (2003)

[Van03b] Vannucci, E.: An Evaluation of the Riskiness of Unit Linked Policies with Minimal Return Guarantees. In: Proceedings of the 6th Spanish-Italian Meeting on Financial Mathematics, vol. II, 569–582 (2003)

Spatial Aggregation in Scenario Tree Reduction

Diana Barro, Elio Canestrelli and Pierangelo Ciurlia

Summary. The solution of a multistage stochastic programming problem needs a suitable representation of uncertainty which may be obtained through a satisfactory scenario tree construction. There is a trade-off between the level of accuracy in the description of the stochastic component and the computational tractability of the resulting scenario-based problem. In order to face such a trade-off which plays a crucial role in the determination of the optimal solution, we discuss methods that allow progressive reductions of a given scenario tree by means of spatial aggregation. In this process it is important to take into account the choice of proper aggregation criteria in order to try to preserve all the relevant information.

Key words: Stochastic Programming; Scenario tree reduction; Spatial Aggregation.
JEL Classification Numbers: C15, C61, C63. MathSci Classification Numbers: 65C05, 65K05.

1 Introduction

One possible way to cope with uncertainty in multi-period stochastic optimization problems is to use scenarios to describe future evolutions for the variables of interest. The introduction of a scenario tree structure to describe uncertainty allows the transformation of a stochastic programming problem into a deterministic equivalent optimization problem.

Although the resulting scenario based problem represents an approximation of the original stochastic problem, it may still be difficult to solve directly due to its large dimensions. To deal with this difficulty we can resort to either proper optimization methods, which exploit the structure of the problem (see among others [BC05a, BC05b]), or aggregation techniques which reduce the dimensions of the original scenario tree.

The construction of a scenario tree is a complex task which requires not only the modelling of the random data and the estimation of the underlying stochastic processes, but also the choice of proper discretization, approximation or sampling

Cira Perna, Marilena Sibillo (Eds.): Mathematical and Statistical Methods in Insurance and Finance

techniques. We focus on the problem of the choice of methods by which a previously generated scenario tree may be optimally reduced to a desired dimension.

Given a scenario tree and its initial probability distribution, our aim is to determine an optimal sequence of reduced trees and probability distributions such that some properties of probability metrics are satisfied. In more detail we are interested in selecting a subset of the original scenario set with a distribution probability which is the closest to the initial one in order to preserve as much of the information content as possible.

The paper is organized as follows. In Section 2 we give a brief overview of the literature on the scenario tree reduction. In Section 3 we present and discuss a new reduction algorithm which introduce two relevant improvements with respect to the method originally proposed in [GKHR03]. Section 4 concludes.

2 Scenario Tree Reduction Using Aggregation Methods

Several recent contributions to the literature address the problem of how to suitably reduce the scenario tree which models the uncertainty in stochastic programs, both from a theoretical point of view (see for example [DGKR03, HR03]) and from a practical one (see for example [GKHR03, K98]).

The optimal reduction problem is strictly related to both the process of scenario tree generation (see [DCW00, P01]), and the analysis of stability properties of optimal solution for stochastic programming problems (see [D90, HRS05, RR02, S00, S94]).

In order to progressively reduce the initial size of a scenario tree we can apply different strategies based on either *spatial aggregation* or *time aggregation*, as well as on several combinations of them.

Time aggregation is obtained by reducing the number of decision stages in the problem. A typical example is given by the multi-period two-stage decision problem which represents a convenient form used to recast multi-stage stochastic programs. In this problem the planning horizon refers to a generic time horizon T while there are only two decision stages; i.e., there is no correspondence between time periods and decision stages. Spatial aggregation consists of a reduction of the scenario tree by deleting scenarios, i.e., by decreasing the number of realizations for the random quantities. These two strategies can be jointly applied to obtain a substantial reduction in the original scenario tree. In [K98] an application of time and state aggregations to built arbitrage-free scenario trees is presented.

In this paper we concentrate on a scenario reduction method based on a particular class of probability metrics. This method allows us to obtain a subset of scenarios of prescribed cardinality or accuracy, from the original scenario tree.

3 A Spatial Aggregation Method for Scenario Tree Reduction

The reduction in the dimension of the scenario tree, used to describe the random data in a stochastic programming problem, is often required in order to be able to solve numerically the resulting optimization problem. In this contribution we focus on a spatial reduction approach proposed by [GKHR03] which is based on the use of probability metrics.

We consider a given scenario tree which describes or even approximates the stochastic component of the problem and we purpose to reduce the dimension of the tree trying to preserve all the relevant information for the optimization problem.

We denote with P the original probability distribution on the scenario tree and with Q the probability distribution on the reduced tree. We assume that both P and Q are discrete probability measures with finitely many realizations (scenarios). In more detail we denote with ξ_i and $p_i = P(\{\xi_i\}),\ i = 1, \dots, N$, the scenarios and their probabilities in the original tree and with ξ_j and $q_j = Q(\{\xi_j\}),\ j = 1, \dots, M$, with $M < N$, the scenarios and their probabilities in the reduced tree.

The optimal scenario reduction problem aims to find the best discrete approximating probability measure Q^* having a prescribed number M of scenarios. The optimality is defined with respect to a measure μ_c of the distance between the two discrete distributions.

According to the notation used in [DGKR03] the reduction problem is defined as

$$\min \left\{ \mu_c \left(\sum_{i=1}^{N} p_i \delta(\xi_i), \sum_{j=1}^{M} q_j \delta(\xi_j) \right) \right\} \tag{1}$$

$$\sum_{j=1}^{M} q_j = 1 \tag{2}$$

$$q_j \geq 0 \ \forall j = 1, \dots, M \tag{3}$$

where $\mu_c(\cdot)$ denotes a proper distance measure among the scenarios and $\delta(\cdot)$ is the Dirac function. For a discussion on the selection of the distance function see [DGKR03, RR02].

Other constraints may be added to problem (1)-(3) in order to ensure desired properties of the probability distribution Q such as moment matching conditions (see for example [HW01]).

Problem (1)-(3) is very difficult to solve for general probability measures P and Q and functions μ_c. Nevertheless if the original probability measure and the approximating one are both discrete with a finite number of scenarios, and if the chosen probability metric belongs to the Kantorovich (or transportation) family of metrics, the problem becomes more tractable, see [DGKR03, RR02].

In more detail, in the following we consider the Kantorivich functional defined as

$$\hat{\mu}_c(P,Q) := \min_{\eta_{ij}} \left\{ \sum_{i=1}^{N} \sum_{j=1}^{M} c_T(\xi_i, \xi_j)\eta_{ij} \right\} \tag{4}$$

$$\sum_{i=1}^{N} \eta_{ij} = q_j \quad \forall j \tag{5}$$

$$\sum_{j=1}^{M} \eta_{ij} = p_i \quad \forall i \tag{6}$$

$$\eta_{ij} \geq 0 \;\; \forall i, j\,, \tag{7}$$

where $c_t(\cdot)$ measures the distance among two scenarios ξ_i and ξ_j and is defined as

$$c_T(\xi_i, \xi_j) := \sum_{t=1}^{T} |\xi_{i,t} - \xi_{j,t}|\,, \tag{8}$$

with $|\cdot|$ a suitable norm in $\mathbb{R}^n$.

Given that the original probability measure P and the approximating one Q are discrete with finitely many scenarios, the Kantorovich distance represents the optimal value of a linear transportation problem where the transport metric measures the cost of moving a scenario ξ_i, or a set of scenarios, of the original tree to the nearest scenario in the new reduced tree.

Moreover, the choice of a probability metric in the Kantorovich family of metrics allows some stability results to be obtained on the optimal values and solutions of the stochastic programs related to the scenario trees. For a more detailed discussion on this topics we refer, for example, to [RR02, DGKR03].

In [GKHR03] the authors propose two heuristics to solve the optimal reduction problem. The first aims at selecting the scenarios that will be deleted (*backward procedure*), while the second one selects the scenarios that will be maintained. In the following, we briefly present the main idea of the first approach and we propose two major improvements in order to overcome some drawbacks of the original approach.

Let $J = \{1, \ldots, N\} \setminus \{1, \ldots, M\}$ be the set of the indexes of deleted scenarios. For a fixed J, the minimal distance can be computed explicitly, see [DGKR03], and is given by

$$\sum_{i \in J} p_i \cdot \min_{j \notin J} c_T(\xi_i, \xi_j)\,. \tag{9}$$

The probability q_{j^*} of the surviving scenarios ξ_i; $j \in \{1, \ldots, M\}$, is computed according to the following optimal redistribution rule:

$$q_{j^*} = p_j + \sum_{i \in J} p_i \quad \forall j \notin J\,, \tag{10}$$

in which the surviving scenario is endowed with the original probability and the probabilities of the deleted scenarios which are close to it according to the chosen metric. For a more detailed discussion on these results see [DGKR03].

In the reduction algorithm proposed by [GKHR03], only the information on the probability of the deleted scenarios survives, through an optimal redistribution rule, while the information related to the values in the deleted nodes disappears.

Starting from the method described in [GKHR03], in this contribution we propose an aggregation algorithm in which we keep trace not only of the probabilities of the deleted scenarios, but also of the values of the deleted nodes. In this way we reduce the amount of information which is lost at each step of the reduction process. Moreover, in order to account not only for the distance among the probability distributions but also between filtrations, we consider a reduction which operates only on portions of the original tree thus allowing to preserve the filtration structure.

The proposed algorithm works for a generic scenario tree built to describe the behavior of random data in a multistage stochastic programming problem. This means that there are no restrictions on the size or structure of the tree that has to be reduced.

We focus on the optimal reduction problem defined as the choice of the less significant scenario, i.e. the scenario that can be deleted with the minimum loss of information. This choice allows us to monitor the reduction process more efficiently. Other approaches can be used, for example, the optimal number of scenarios to be deleted can be computed at each step of the algorithm, see [HR03].

According to the previously used notation, let $\{\xi_t\}_{t=1}^T$ be n-dimensional stochastic process with parameter set $\{1, \ldots, T\}$ and ξ^i, ξ^j denote scenarios.

Moreover, to describe the tree structure we denote with $k_t = K_{t-1} + 1, \ldots, K_t$, for $t = 1, \ldots, T$ and with $K_0 = 0$ and $K_1 = 1$, the indexes of nodes at time t, and with $S = K_T - K_{T-1}$ the number of scenarios in the original event tree, i.e., the set of scenarios is $\Omega := \{1, \ldots, s, \ldots, S\}$. Let $\mathcal{D}(k_t)$, for $t = 1, \ldots, T$, be the index set of scenarios having k_t as ancestor node at time t, i.e., $\mathcal{D}(k_t) \subseteq \Omega$, and $\pi_{k_{t,s}}^{k_{t+1,s}}$ be the probability of transition from node k_t at time t to its descending node k_{t+1} at time $t + 1$ for scenario s, i.e., $\pi_{k_{t,s}}^{k_{t+1,s}} \geq 0, \forall\, k_t,\ t = 1, \ldots, T - 1$. The probability of each scenario is given by $p_s := \prod_{t=1}^{T-1} \pi_{k_{t,s}}^{k_{t+1,s}}$, for $s = 1, \ldots, S$. Let J denote the index set of deleted scenarios and $c_T\left(\xi^i, \xi^j\right)$ distance between scenarios $\left\{\xi_t^i\right\}_{t=1}^T$ and $\left\{\xi_t^j\right\}_{t=1}^T$, with $i, j \in \mathcal{D}(k_t)$.

The *information structure* in the original tree is fully described by a sequence $\left\{\mathcal{P}_t\right\}_{t=1}^T$ of partitions of the sample space Ξ, with $\mathcal{P}_t = \left\{\mathcal{D}(K_{t-1} + 1), \ldots, \mathcal{D}(k_t), \ldots, \mathcal{D}(K_t)\right\}$ and satisfying the property that each $\mathcal{D}(k_t) \in \mathcal{P}_t$ is equal to the union of some elements in $\mathcal{P}_{t+1}$, $\forall\, t, t = 1, \ldots, T - 1$.

The *optimal reduction problem* can be described as follows. For each discrete-time horizon $\{t, \ldots, T\}$, with $t = T-1, T-2, \ldots, 1$, we calculate

$$\min_{k_t \in \{K_{t-1}+1, \ldots, K_t\}} \left\{ \min_{i \in \mathcal{D}(k_t)\setminus J} \left[p_i \cdot \min_{\substack{j \neq i \\ j \in \mathcal{D}(k_t)\setminus J}} c_T\left(\xi^i, \xi^j\right) \right] \right\}, \tag{11}$$

where

$$c_T\left(\xi^i, \xi^j\right) := \sum_{\tau=1}^{T} |\xi_\tau^i - \xi_\tau^j| \tag{12}$$

and $|\cdot|$ denotes some norm on $\mathbb{R}^n$, i.e., $c_T(\cdot,\cdot)$ measures the distance between a pair of scenarios on the whole time horizon $\{1, \ldots, T\}$.

If the minimum is attained at k_t^*, then the scenario i^* is deleted and

$$j^* \in \arg \min_{j \neq i^*} c_T\left(\xi^{i^*} \xi^j\right), \tag{13}$$

is the aggregating scenario, with $i^*, j^* \in \mathcal{D}(k_t^*)$.

The *optimal redistribution rule* for the probabilities is defined, according to the results in [GKHR03], as follows:

$$\tilde{\pi}_{k_\tau, j^*}^{k_{\tau+1}, j^*} = \pi_{k_\tau, j^*}^{k_{\tau+1}, j^*} + \pi_{k_\tau, i^*}^{k_{\tau+1}, i^*}, \tag{14}$$

for $\tau = t, \ldots, T-1$.

Beside the aggregation on the probabilities, we introduce the following aggregation step to take into account the values of the deleted scenario

$$\tilde{\xi}_{\tau+1}^{j^*} = \frac{\pi_{k_\tau, j^*}^{k_{\tau+1}, j^*} \xi_{\tau+1}^{j^*} + \pi_{k_\tau, i^*}^{k_{\tau+1}, i^*} \xi_{\tau+1}^{i^*}}{\pi_{k_\tau, j^*}^{k_{\tau+1}, j^*} + \pi_{k_\tau, i^*}^{k_{\tau+1}, i^*}}, \tag{15}$$

i.e., the values of the aggregating scenario is computed as a weighted average of the values of the original scenario and the deleted one, where the weights are given by the corresponding probabilities.

In the following we describe in more detail the algorithm proposed for the optimal reduction problem.

Algorithm 1 *Optimal reduction algorithm*

Require: A previously generated event tree with S scenarios
Ensure: At each iteration a scenario $i^* \in \mathcal{D}(k_t^*)$ is deleted

for $t = T - 1$ to 1 **do**
  while $\text{Card}(\mathcal{D}(k_t)) = 1, \forall k_t$, **do**
    Calculate the distances of scenarios pairs for each element of partition $\mathcal{P}_t$

$$c_{ij} := c_T(\xi^i, \xi^j), \quad i, j \in \mathcal{D}(k_t) \setminus J, \ \forall \mathcal{D}(k_t) \in \mathcal{P}_t$$

    Sort the records in matrix form

$$c(k_t) := \left[c_{ij} \ ; i, j \in \mathcal{D}(k_t) \setminus J\right], \ k_t = K_{t-1} + 1, \ldots, K_t$$

    for all k_t **do**
      Calculate

$$c_{ii}^{k_t} := \min_{\substack{j \neq i \\ j \in \mathcal{D}(k_t) \setminus J}} c_{ij}, \quad i \in \mathcal{D}(k_t)$$

and

$$z_i^{k_t} := p_i \, c_{ii}^{k_t}, \quad i \in \mathcal{D}(k_t)$$

      Choose

$$\tilde{i}(k_t) \in \arg \min_{i \in \mathcal{D}(k_t)} z_i^{k_t}$$

    end for
    Sort the records in matrix form

$$Z(t) := \left[z_{\tilde{i}(k_t)}^{k_t} \ ; \ k_t = K_{t-1} + 1, \ldots, K_t\right]$$

    Choose

$$k_t^* \in \arg \min_{k_t \in \{K_{t-1}+1, \ldots, K_t\}} Z(t)$$

    Set

$$J := \{i^*\}$$

where

$$i^* := \tilde{i}(k_t^*)$$

  end while
end for

4 Concluding Remarks

In this contribution we briefly discuss the problem of optimal scenario tree reduction for multistage stochastic programming problems and we propose two major improvements to a backward reduction algorithm proposed by [GKHR03]. Our modified algorithm allows us to take into account not only the transition probabilities but also the values of the deleted scenarios and to aggregate them in an optimal way. Moreover, it is designed to handle general tree structure since it preserve the information structure of the tree.

References

[BC05a] Barro, D. and Canestrelli, E.: Dynamic portfolio optimization: time decomposition using the Maximum Principle with a scenario approach. European Journal of Operational Research, **163**, 217–229 (2005a)

[BC05b] Barro, D. and Canestrelli, E.: A decomposition approach in multistage stochastic programming. Rendiconti per gli Studi Economici Quantitativi, 73–88 (2005b)

[D90] Dupačová, J.: Stability and sensitivity analysis for stochastic programming. Annals of Operations Research, **27**, 21–38 (1990)

[DCW00] Dupačová, J. and Consigli, G. and Wallace, S. W.: Scenarios for multistage stochastic programs. Annals of Operations Research, **100**, 25–53 (2000)

[DGKR03] Dupačová, J. and Gröwe-Kuska, N. and Römisch, W.: Scenario reduction in stochastic programming, an approach using probability metrics. Mathematical Programming Series A., **95**, 493–511 (2003)

[GKHR03] Gröwe-Kuska, N. and Heitsch, H. and Römisch, W.: Scenario reduction and scenario tree construction for power management problems. IEEE Bologna Power Tech Proceedings (A. Borghetti, C. A. Nucci, M. Paolone eds). (2003)

[HR03] Heitsch, H. and Römisch, W.: Scenario reduction algorithms in stochastic programming. Computational Optimization and Applications, **24**, 187–206 (2003)

[HRS05] Heitsch, H. and Römisch, W. and Strugarek C.: Stability of multistage stochastic programming. Stochastic programming e-prints, **16** (2005)

[HW01] Høyland, K. and Wallace S. W.: Generating scenario trees for multi-stage decision problems. Management Science, **47**, 295–307 (2001)

[K98] Klaassen, P.: Financial Asset-Pricing theory and stochastic programming models for asset/liability management: a synthesis. Management Science, **44**, 31–48 (1998)

[P01] Pflug, G. Ch.: Scenario tree generation for multiperiod financial optimization by optimal discretization. Mathematical Programming Series B., **89**, 251–271 (2001)

[RR02] Rachev, S. T and Römisch W.: Quantitative stability in stochastic programming: The method of probability metrics. Mathematics of Operations Research, **27**, 792–818 (2002)

[S00] Schultz, R.: Some aspects of stability in stochastic programming. Annals of Operations Research, **100**, 55–84 (2000)

[S94] Shapiro, A.: Quantitative stability in stochastic programming. Mathematical Programming, **67**, 99–108 (1994)

Scaling Laws in Stock Markets. An Analysis of Prices and Volumes

Sergio Bianchi and Augusto Pianese

Summary. The scaling behaviour of both log-price and volume is analyzed for three stock indexes. The traditional approach, mainly consisting of the evaluation of particular moments such as variance or higher absolute moments, is replaced by a new technique which allows the estimation of the self-similarity parameter on the whole empirical distribution designed by any time horizon. In this way, the method we propose attaches its own scaling parameter to any two given time lags, so defining a scaling surface whose properties give information about the nature of the analyzed process. We conclude that, for the log-price process, self-similarity is rejected with a frequency much larger than that assumed by the confidence interval and, when not rejected, the scaling parameter heavily changes with the considered pair of time horizons. Opposite evidence is provided for the volumes, characterized by (generally low) self-similarity parameters which are somewhat uniform with respect to the pairs of time horizons.

Key words: Scaling; Self-similarity; Stock indexes.

1 Introduction

Since the pioneering work by [MAN63], the scaling properties of financial returns have become the subject of a growing number of contributions for both theoretical and practical reasons. From a theoretical viewpoint, scaling invariance implies the absence of preferred investment time horizons and the consequent universality and parsimony of the model; practical advantages concern the possibility of building models stable under aggregation, analytically simple and governed by a small number of parameters. Empirical evidence of multi-scaling in finance is provided by [MDO90], [CL93], [FLL94], [SSL00], [VA98], [KL01] and [GX03] for the FX markets; by [DAD05], [GPA99], [AI02], [BER01], [ONA04] and [YMH05] for the stock markets and by [AMS98] and [MDB00] for the future markets. Scaling analysis has also been employed to probe the different degree of market development (see e.g. [DIM06]). More recently, volatility scaling has become the focus of several works following the Basel Accords, which have reaffirmed the square–root–of–time rule (correct only under self-similarity with parameter 1/2) as a proxy for estimating volatility on different time horizons see [MEN04], [DZ04] and, ante litteram,

Cira Perna, Marilena Sibillo (Eds.): Mathematical and Statistical Methods in Insurance and Finance

[DHI97] for an analysis of the drawbacks of this rule of thumb. Using different techniques, basically two types of scaling behaviours are studied: the scaling of some volatility measures (typically variance or absolute moments of the returns) as a function of the time interval and, once this has been fixed, the scaling of the tails of the distribution of returns as a function of the size of the variation [DGM01]. To characterize the scaling properties of financial markets, empirical tests generally use the rescaled range analysis (introduced by [HUR51] and modified by [LO91]), the multi-affine analysis [PBH94], the more recent Detrended Fluctuation Analysis [PBH94], the ARFIMA estimation by exact maximum likelihood, the moving average-like analysis methods and the Average Wavelet Coefficient Method (see, e.g. [GSW01]). In this paper we analyze the scaling behaviour of prices and volumes for three stock indexes representative of the U.S. (Dow-Jones Industrial Average), Europe (Footsie 100) and Asia (Hang Seng). The interest towards the traded volumes is suggested by the Multifractal Model of Asset Returns (MMAR) which, introduced by [CF02], is the focus of a wide debate.

The basic idea beyond the MMAR is to compound a Brownian motion (eventually a fractional one) with a multifractal trading time, defined as the cumulative distribution function of a self-similar multifractal measure which deforms the physical time in order to take into account the different number of transactions per unit of time. According to this model, we expect to find self-similarity in the traded volume. Our analysis is performed using for the very first time a new class of estimators of the self-similarity parameter introduced by [BA04]. The method, which is distribution-based but distribution-free, provides a very robust and immediate representation of the scaling relation between time horizons.

2 Self Similarity and Scaling

2.1 Theoretical background

Let us shortly recall the basic definition of (strong) self-similarity which will be useful later.

Definition 1. *The continuous time, real-valued process* $\{X(t),\ t \in T\}$, *with* $X(0) = 0$, *is self-similar with index* $H_0 > 0$ *(concisely,* H_0*-ss) if, for any* $a \in \mathbb{R}^+$ *and any integer* k *such that* $t_1, \ldots, t_k \in T$, *the following equality holds for its finite-dimensional distributions*

$$\{X(at_1), X(at_2), \ldots, X(at_k)\} \stackrel{d}{=} \{a^{H_0}X(t_1), a^{H_0}X(t_2), \ldots, a^{H_0}X(t_k)\}. \quad (1)$$

Equality (1) implies

$$\mathbb{E}(|X(t)|^q) = t^{H_0 q}\mathbb{E}(|X(1)|^q), \quad (2)$$

which allows the testing of self-similarity by the scaling of the sample (absolute) moments of $X(t)$. With respect to the definition of self-similarity, this approach leads to weaker conclusions because the reverse implication (from (2) to (1)) is not necessarily true. The problem is addressed by [BA04], who reformulates relation (1) in an equivalent way by introducing a proper metric on the space of the rescaled probability distribution functions (pdf's) as follows. Let $\mathcal{A}$ be any bounded subset of $\mathbb{R}^+$, $\mathfrak{a} = \min(\mathcal{A})$ and $\mathfrak{A} = \max(\mathcal{A}) < \infty$, for any $a \in \mathcal{A}$, consider the k-dimensional distribution Φ of the a-lagged process $X(at)$. Equality (1) becomes

$$\Phi_{\mathbb{X}(a)}(\mathbf{x}) = \Phi_{a^{H_0}\mathbb{X}(1)}(\mathbf{x}), \tag{3}$$

where $\mathbb{X}(a) = (X(at_1), \dots, X(at_k))$ and $\mathbf{x} = (x_1, \dots, x_k) \in \mathbb{R}^k$. Introducing the variable H, for a self-similar process, it follows that

$$\begin{aligned}\Phi_{a^{-H}\mathbb{X}(a)}(x) &= \Pr\left(a^{-H}X(at_1) < x_1, \dots, a^{-H}X(at_k) < x_k\right) = \text{by } H_0\text{-ss} =\\ &= \Pr\left(a^{H_0-H}X(t_1) < x_1, \dots, a^{H_0-H}X(t_k) < x_k\right) = \Phi_{a^{H_0-H}\mathbb{X}(1)}(x) = \\ &= \Pr\left(X(t_1) < a^{H-H_0}x_1, \dots, X(t_k) < a^{H-H_0}x_k\right) = \Phi_{\mathbb{X}(1)}(a^{H-H_0}x).\end{aligned} \tag{4}$$

Denoted by ρ the distance function induced by the sup-norm $\|\cdot\|_\infty$ on the space Ψ_H of the k-dimensional pdf's of $\{a^{-H}X(at)\}$ with respect to the set $\mathcal{A}$, the diameter of the metric space (Ψ_H, ρ)

$$\delta^k(\Psi_H) = \sup_{\mathbf{x}\in\mathbb{R}^k} \sup_{a_i, a_j \in \mathcal{A}} \left|\Phi_{a_i^{-H}\mathbb{X}(a_i)}(\mathbf{x}) - \Phi_{a_j^{-H}\mathbb{X}(a_j)}(\mathbf{x})\right| \tag{5}$$

measures the *discrepancy* among the distributions of the rescaled process. If self-similarity holds, then for any $H \neq H_0$[1], by (4), one can trivially notice that being Φ a distribution

$$\sup_{a_i, a_j \in \mathcal{A}} \left|\Phi_{a_i^{-H}\mathbb{X}(a_i)}(\mathbf{x}) - \Phi_{a_j^{-H}\mathbb{X}(a_j)}(\mathbf{x})\right| = \left|\Phi_{\mathfrak{a}^{-H}\mathbb{X}(\mathfrak{a})}(\mathbf{x}) - \Phi_{\mathfrak{A}^{-H}\mathbb{X}(\mathfrak{A})}(\mathbf{x})\right| \tag{6}$$

and equation (5) reduces itself to the well known statistics of Kolmogorov-Smirnov, enriching the self-similarity analysis with an inferential support.

For a truly H_0-ss process, (5) has been proven to be non-increasing for $H < H_0$ and non-decreasing for $H > H_0$; combining this result with the uniqueness of the self-similarity index ([LAM62]), H_0 is estimated as

$$H_0 = \arg\min_{H\in(0,1)} \sup_{\mathbf{x}\in\mathbb{R}^k} \left|\Phi_{\mathfrak{a}^{-H}\mathbb{X}(\mathfrak{a})}(\mathbf{x}) - \Phi_{\mathfrak{A}^{-H}\mathbb{X}(\mathfrak{A})}(\mathbf{x})\right|,$$

provided that $\delta^k(\Psi_{H_0})$ is not significant once the confidence interval has been fixed.

---

[1] For a H_0-ss process, when $H = H_0$, it is trivial to check that
$\sup_{a_i, a_j \in \mathcal{A}} \left|\Phi_{a_i^{-H}\mathbb{X}(a_i)}(\mathbf{x}) - \Phi_{a_j^{-H}\mathbb{X}(a_j)}(\mathbf{x})\right|$ collapses to zero, whatever a_i and a_j.

2.2 Scaling surfaces

Since for a self-similar process from (3), it follows that

$$\Phi_{a^{-H_0}\mathbb{X}(a)}(\mathbf{x}) = \Phi_{b^{-H_0}\mathbb{X}(b)}(\mathbf{x}),$$

where a and b denote any two time horizons, the diameter (5) can be calculated with respect to any pair of lags $a < b$ and the more general self-similarity parameter can be written as

$$H_0(a,b) = \underset{H\in(0,1)}{\arg\min} \sup_{\mathbf{x}\in\mathbb{R}^k} \left|\Phi_{b^{-H}\mathbb{X}(b)}(\mathbf{x}) - \Phi_{a^{-H}\mathbb{X}(a)}(\mathbf{x})\right|. \tag{7}$$

The last relation represents a useful form for testing the scaling properties of a time series by means of the pairwise comparisons of the time horizons (a, b). From a geometrical viewpoint, the idea consists of associating at each pair (a, b) with $a < b$ for $a, b \in \mathcal{A}$ the third coordinate given by the estimated self-similarity parameter $H_0(a, b)$ whenever the null hypothesis of the Kolmogorov-Smirnov test (identity of the rescaled distributions $\Phi_{b^{-H}\mathbb{X}(b)}(\mathbf{x})$ and $\Phi_{a^{-H}\mathbb{X}(a)}(\mathbf{x})$) is not rejected at a given p-level.

In an equivalent way, we can see the result in the form of a strictly lower triangular matrix displaying the maximum horizons on the rows, and the minimum horizons on the columns, whose elements are the parameters H_0's.

Basically, in our approach the definition of self-similarity is not required to hold for any time horizon; we will be satisfied to determine which (if any) parameter H_0 minimizes the distance (7) to such an extent to be statistically negligible. Therefore, once set $\mathcal{A}$ has been fixed, analysis of the shape of the surface drawn by points (a, b, H_0) will provide information about the nature of the time series: a H_0-high plane will indicate that the process is truly H_0-ss (relatively to set $\mathcal{A}$); on the contrary, a jagged surface is expected to be generated by a multi-scaling process. As an example, Fig. 1 reproduces the scaling surface of the increments of a standard Brownian motion, which are known to be self-similar with parameter $\frac{1}{2}$. Indeed, the surface is regular and $\frac{1}{2}$-high.

3 Empirical Application

3.1 Dataset and methodology

The methodology described above has been applied to the three main stock indexes summarized in Table 1. The different sizes of the number of observations is due to the availability of the traded volumes.

For each time series the daily log-price variations have been calculated and considered as input for the scaling analysis. The lags a and b have been taken in the set of time horizons $\mathcal{A} = \{1, 2, \ldots, 125\}$, corresponding to about six months of trading days, and the null hypothesis has been tested for $\alpha = 0.05$ and $\alpha = 0.01$. The scaling surfaces are obtained by filtering only those estimated H_0's for which the corresponding diameter (7) is below the critical value of the Smirnov statistics.

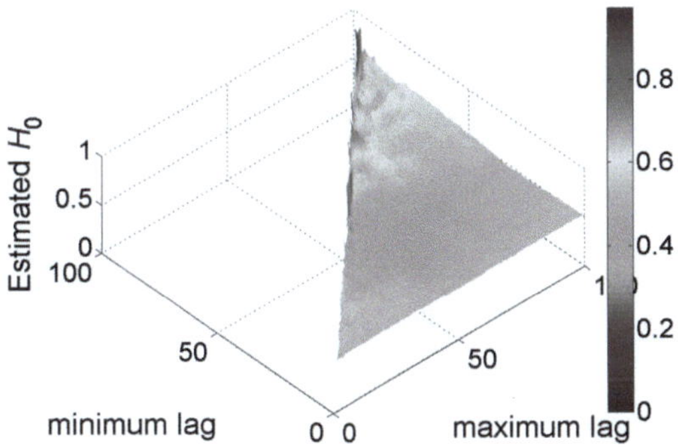

Fig. 1. Scaling surface of an sBm.

Table 1. Analyzed dataset.

Index	# obs	Starting day	Ending day
Dow-Jones Ind. Av.	5,000	11/04/86	08/29/06
Footsie 100	944	12/03/02	08/29/06
Hang Seng	1,225	09/19/02	08/29/06

3.2 Discussion of result

Fig. 2, 3 and 4, displaying the XY projection of the scaling surfaces, summarize the results of our analysis applied to returns (panels (a)'s) and volumes (panels (b)'s) of the three stock indexes. The holes of the scaling surfaces denote that the values of the corresponding diameter (7) exceed the critical threshold, given the confidence level α. The results, similar among the three indexes, are strongly different in terms of returns and volumes and deserve a separate discussion.

Price: for all three time series, self-similarity is rejected with a frequency much larger than that assumed by the confidence interval. Moreover, when not rejected, the parameter H_0 heavily changes with the considered pair of time horizons in almost all the range (0, 1). Some small areas where the parameter seems to be more stable can be distinguished. This behaviour is typical of multifractal processes, characterized by a plethora of scaling exponents.

Volume: unlike the log-price variations, self-similarity is rejected with a frequency which is only slightly larger than that assumed by the confidence interval and, what looks relevant, when not rejected, the parameter H_0 is somewhat constant with respect to the considered pair of time horizons with generally very low values. This result looks strongly consistent with the recently proposed Multifractal Model of Asset Returns (MMAR), which assumes a multifractal self-similar measure to model the number of shares traded per unit of physical time. The constancy we observe in

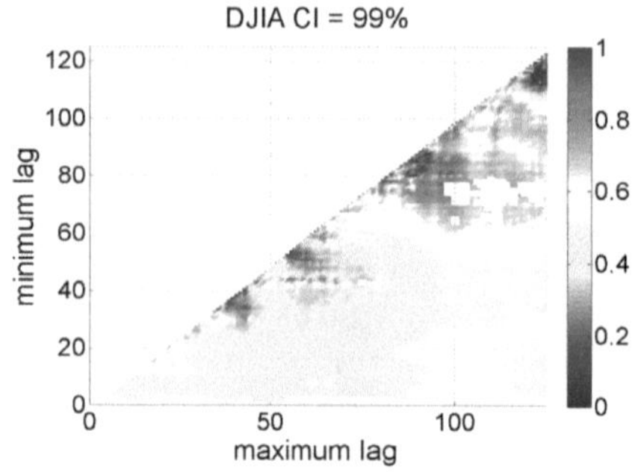

Fig. 2a

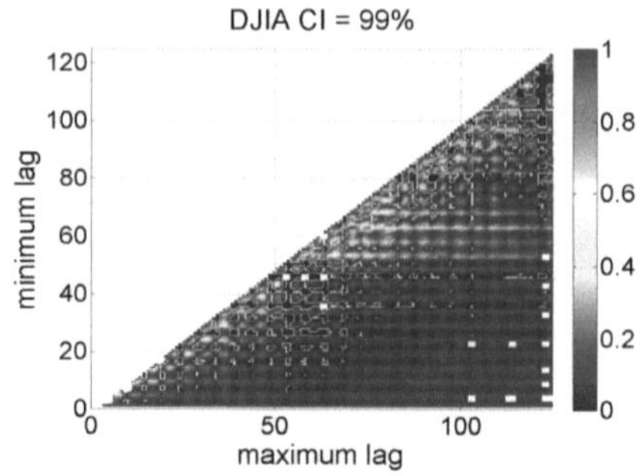

Fig. 2b

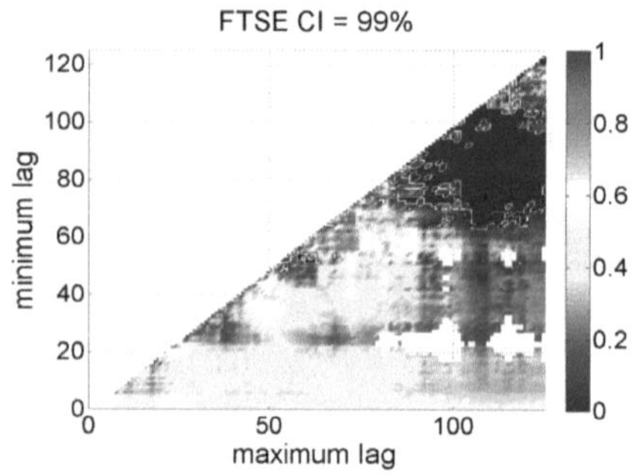

Fig. 3a

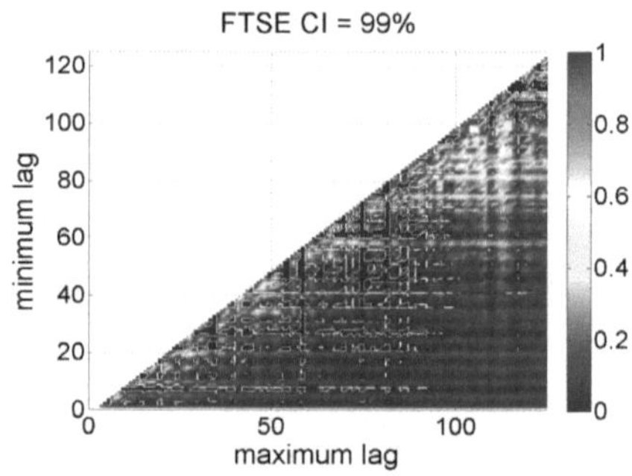

Fig. 3b

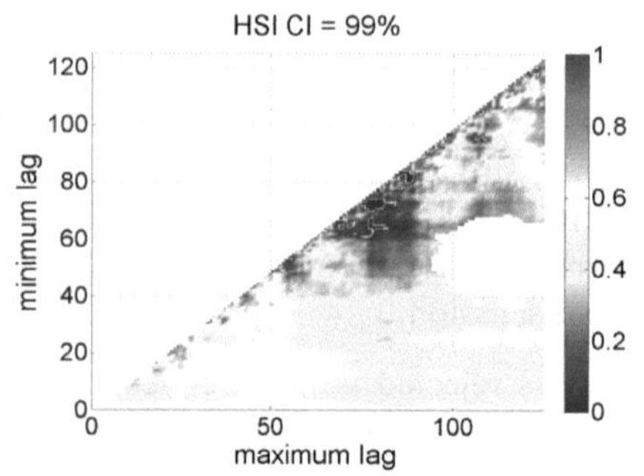

Fig. 4a

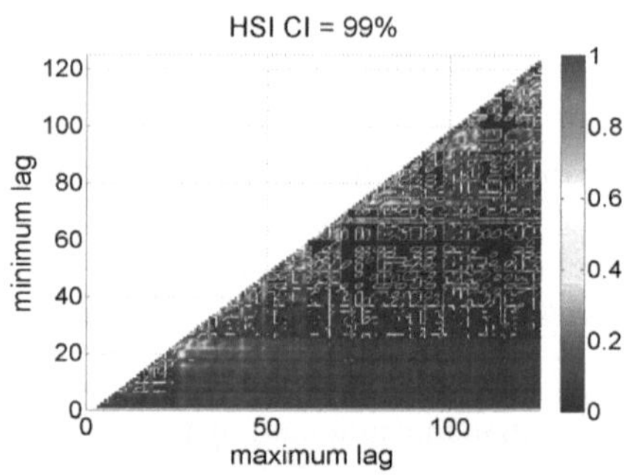

Fig. 4b

the self-similarity parameter with respect to the pairwise comparisons of the time horizons is coherent with what the MMAR should produce.

4 Conclusion and Further Developments

Applying for the very first time a new technique for estimating the self-similarity parameter, evidence has been provided that the scaling behaviour of three major financial indexes is much more complex than the one stated by the traditional theory. More precisely, completely different rules seem to be followed by prices and volumes. The former are governed by a number of scaling exponents without any apparent substantial regularity while the scaling exponents of the latter look somewhat regular and almost constant with very low values. These findings agree with the MMAR and prove themselves very useful in a variety of situations. For example, the values $H_0(a, b)'s$ can be used to improve the estimation of the annualized volatility, usually calculated by means of the well known (and wrong when applied to actual financial data) square–root–of–time rule. The improvement can be easily obtained by substituting the constant exponent $H_0 = 1/2$ with the parameter $H_0(1, b)$, where b represents the time horizon the volatility is "annualized" for. This issue deserves further analysis because of its relevance for risk management.

References

[AMS98] Arneodo A., J. F. Muzy and D. Sornette, *Direct causal cascade in the stock market*, The European physical Journal B 2, 277–282 (1998)

[AI02] Ausloos M. and K. Ivanova, *Multifractal nature of stock exchange prices*, Computer Physics Communications 147: 582–585 (2002)

[BER01] Bershadskii A., *Invasion–percolation and statistics of US Treasury bonds*, Physica A: Statistical Mechanics and its Applications, 300, 3-4 , 539–550 (2001)

[BA04] Bianchi S., *A new distribution-based test of self-similarity*, Fractals, 12, 3, 331–346 (2004)

[CF02] Calvet L. and Fisher A., *Multifractality in Asset Returns: Theory and Evidence*, Review of Economics and Statistics 84, 381–406 (2002)

[CL93] Cheung Y. and K. Lai, *Do gold market returns have long memory?*, Financial Review 28, 181–202 (1993)

[DGM01] Dacorogna M. M., Genay R., Müller U., Olsen R. B., Pictet O.V., An Introduction to High-Frequency Finance, Academic Press, London (2001)

[DZ04] Daníelsson J., Zigrand J. P., *On time-scaling of risk and the square-root-of-time rule*, FMG Discussion Paper dp439, London School of Economics and Political Science, London (2004)

[DHI97] Diebold F. X., Hickman A., Inoue A., Schuermann T., *Converting 1-Day Volatility to h-Day Volatility: Scaling by is Worse than You Think*, Penn Institute for Economic Research Working Papers, 97-030 (1997)

[DIM06] Di Matteo T., *Multi-Scaling in Finance*, to appear in Quantitative Finance (2006)

[DAD05] Di Matteo T., T. Aste and M. M. Dacorogna, *Long-term memories of developed and emerging markets: Using the scaling analysis to characterize their stage of development*, Journal of Banking & Finance, 29, 827–851 (2005)

[FLL94] Fang H., K. S. Lai and M. Lai, *Fractal structure in currencyfutures price dynamics*, Journal of Futures Markets, 14, 2, 169–181 (1994)

[GPA99] Gopikrishnan P., Plerou V., Amaral L. A. N., Meyer M. and Stanley H. E., *Scaling of the distribution of fluctuations of financial market indices*, Phys. Rev. E Stat. Nonlin. Soft Matter Phys. 60, 5, 5305–5316 (1999)

[GSW01] Gençay, R., F. Seluuk and B. Whitcher, *Scaling properties of foreign exchange volatility*, Physica A 289, 249–266 (2001)

[GX03] Gençay R. and Xu Z., *Scaling, self-similarity and multifractality in FX markets*, Physica A, 323, 578–590 (2003)

[HUR51] Hurst H. E., *Long-term storage capacity of reservoirs*, Transactions of the American Society of Civil Engineers, 116: 770–799 (1951)

[KL01] Karuppiah J. and Los C. A., *Wavelet multiresolution analysis of high-frequency Asian FX rates*, Summer 1997, AFA 2001 New Orleans; Adelaide University, School of Economics Working Paper No. 00-6 (2001)

[LAM62] Lamperti J. W., *Semi-stable processes*, Trans. Amer. Math. Soc., 104, 62–78 (1962)

[LO91] Lo A. W., *Long-Term Memory in Stock Market Prices*, Econometrica, 59, 1279–1313 (1991)

[MAN63] Mandelbrot B. B., *The variation of certain speculative prices*, Journal of Business, 36, 392–417 (1963)

[MEN04] Menkens O., *Value at Risk and Self-Similarity*, Preprint, Centre for Computational Finance and Economic Agents, Department of Economics, University of Essex (2004)

[MDO90] Müller U. A., M. M. Dacorogna, R. B. Olsen, O. V. Pictet, M. Schwartz and C. Morgenegg, *Statistical Study of Foreign Exchange Rates, Empirical Evidence of a Price Change Scaling Law and Intraday Analysis*, Journal of Banking and Finance, 14, 1189–1208 (1990)

[MDB00] Muzy J., J. Delour, and E. Bacry, *Modeling fluctuations of financial time series: From cascade processes to stochastic volatility models*, Eur. J. Phys. B, 17, 537–548 (2000)

[ONA04] Önalan Ö., *Self-similarity and multifractality in financial asset returns*, Computational Finance and its Applications, M. Costantino and C. A. Brebbia (Editors), WIT Press (2004)

[PBH94] Peng C. K., Buldyrev S.V., Havlin S., Simons M., Stanley H. E. and Goldberger A. L., *Mosaic organization of DNA nucleotides*, Phys. Rev. E Stat. Nonlin. Soft Matter Phys. 6849, 1685–1689 (1994)

[SSL00] Schmitt F., D. Schertzer, S. Lovejoy, *Mulitfractal Fluctuations in Finance*, International Journal of Theoretical and Applied Finance, 3, 361–364 (2000)

[VA98] Vandewalle N. and M. Ausloos, *Multi-affine analysis of typical currency exchange rates*, The European Physical Journal B, 4, 2, 257–261 (1998)

[YMH05] Yamasaki K., Muchnik L., Havlin S., Bunde A., Stanley H. E., *Scaling and memory in volatility return intervals in financial markets*, PNAS, (102) 26, 9424–9428 (2005)

Bounds for Concave Distortion Risk Measures for Sums of Risks*

Antonella Campana and Paola Ferretti

Summary. In this paper we consider the problem of studying the gap between bounds of risk measures of sums of non-independent random variables. Owing to the choice of the context of where to set the problem, namely that of distortion risk measures, we first deduce an explicit formula for the risk measure of a discrete risk by referring to its writing as sum of layers. Then, we examine the case of sums of discrete risks with identical distribution. Upper and lower bounds for risk measures of sums of risks are presented in the case of concave distortion functions. Finally, the attention is devoted to the analysis of the gap between risk measures of upper and lower bounds, with the aim of optimizing it.

Key words: Distortion risk measures; Discrete risks; Concave risk measures; Upper and lower bounds; Gap between bounds.
M.S.C. classification: 62H20; 60E15.
J.E.L. classification: D810.

1 Introduction

Recently in actuarial literature, the study of the impact of dependence among risks has become a major and flourishing topic: even if in traditional risk theory, individual risks have usually been assumed to be independent, this assumption is very convenient for tractability but it is not generally realistic. Think for example of the aggregate claim amount in which any random variable represents the individual claim size of an insurer's risk portfolio. When the risk is represented by residential dwellings exposed to danger of an earthquake in a given location or by adjoining buildings in fire insurance, it is unrealistic to state that individual risks are not correlated, because they are subject to the same claim causing mechanism. Several notions of dependence were introduced in the literature to model the fact that larger values of one of the components of a multivariate risk tend to be associated with larger values of the others. In financial or actuarial situations one often encounters random variables of

* This research was partially supported by MIUR.

Cira Perna, Marilena Sibillo (Eds.): Mathematical and Statistical Methods in Insurance and Finance

the type

$$S = \sum_{i=1}^{n} X_i$$

where the terms X_i are not mutually independent and the multivariate distribution function of the random vector $\mathbf{X} = (X_1, X_2, \ldots, X_n)$ is not completely specified but one only knows the marginal distribution functions of the risks. To be able to make decisions, in such cases it may be helpful to determine approximations for the distribution of S, namely upper and lower bounds for risk measures of the sum of risks S in such a way that it is possible to consider a riskiest portfolio and a safest portfolio, where riskiness and safety are both evaluated in terms of risk measures.

With the aim of studying the gap between the riskiest and the safest portfolio, the present contribution addresses the analysis to a particular class of risk measures, namely that of distortion risk measures introduced by Wang [W96]. In this class, the risk measure of a non-negative real valued random variable X is written in the following way:

$$W_g(X) = \int_0^\infty g(H_X(x))dx ,$$

where the distortion function g is defined as a non-decreasing function $g : [0, 1] \to [0, 1]$ such that $g(0) = 0$ and $g(1) = 1$.

Given the choice of this context, it is possible to write an explicit formula for the risk measure of a discrete risk by referring to its writing as sum of layers (Campana and Ferretti [CF05]). Starting from this result, the attention is therefore devoted to the study of bounds of sums of risks in the case of discrete identically distributed random variables. Now the key role is played by the choice of the framework where to set the study: by referring to concave distortion risk measures, in fact, it is possible to characterize the riskiest portfolio where the multivariate distribution refers to mutually comonotonic risks and the safest portfolio where the multivariate distribution is that of mutually exclusive risks. Again, starting from the representation of risks as sums of layers, it is possible to derive explicit formulas for risk measures of upper and lower bounds of sums of risks. The attention is then devoted to the study of the difference between risk measures of upper and lower bounds, with the aim of obtaining some information on random variables for which the gap is maximum or minimum.

The paper is organized as follows. In Section 2, we first review some basic settings for describing the problem of measuring a risk and we recall some definitions and preliminary results in the field of distortion risk measures. We then propose the study of the case of a discrete risk with finitely many mass points in such a way that it is possible to give an explicit formula for its distortion risk measure. Section 3 is devoted to the problem of detecting upper and lower bounds for sums of not mutually independent risks. We present the study of the case of sums of discrete and identically distributed risks in order to obtain upper and lower bounds for concave distortion measures of aggregate claims of the portfolio. Then in Section 4 the attention is focused on the problem of characterizing risks for which the gap between bounds of risk measures is maximum or minimum. Some concluding remarks in Section 5 end the paper.

2 The Class of Distortion Risk Measures

As it is well known, an insurance risk is defined as a non-negative real-valued random variable X defined on some probability space.

Here we consider a set Γ of risks with bounded support $[0, c]$. For each risk $X \in \Gamma$ we denote by H_X its tail function, i.e. $H_X(x) = Pr[X > x]$, for all $x \geq 0$.

A risk measure is defined as a mapping from the set of random variables, namely losses or payments, to the set of real numbers. In actuarial science common risk measures are premium principles; other risk measures are used for determining provisions and capital requirements of an insurer in order to avoid insolvency (see for example Dhaene et al. [DVTGKV04]).

In this paper we consider the distortion risk measure introduced by Wang [W96]:

$$W_g(X) = \int_0^{\infty} g(H_X(x))dx \,, \tag{1}$$

where the distortion function g is defined as a non-decreasing function $g : [0, 1] \to [0, 1]$ such that $g(0) = 0$ and $g(1) = 1$. As it is well known, the *quantile risk measure* and the *Tail Value-at-Risk* are examples of risk measures belonging to this class. In the particular case of a power g function, i.e. $g(x) = x^{1/\rho}$, $\rho \geq 1$, the corresponding risk measure is the *PH-transform risk measure* proposed by Wang [W95].
Distortion risk measures satisfy the following properties (see Wang [W96] and Dhaene et al. [DVTGKV04]):

P1. Additivity for comonotonic risks

$$W_g(S^c) = \sum_{i=1}^{n} W_g(X_i)\,, \tag{2}$$

where S^c is the sum of the components of the random vector $\mathbf{X^c}$ with the same marginal distributions of $\mathbf{X}$ and with the comonotonic dependence structure.

P2. Positive homogeneity

$$W_g(a\,X) = a\,W_g(X) \quad \text{for any non-negative constant } a; \tag{3}$$

P3. Translation invariance

$$W_g(X + b) = W_g(X) + b \quad \text{for any constant } b; \tag{4}$$

P4. Monotonicity

$$W_g(X) \leq W_g(Y) \tag{5}$$

for any two random variables X and Y where $X \leq Y$ with probability 1.

2.1 Discrete risks with finitely many mass points

In the particular case of a discrete risk $X \in \Gamma$ with finitely many mass points it is possible to deduce an explicit formula of the distortion risk measure $W_g(X)$ of X. The key point relies on the fact that each risk $X \in \Gamma$ can be written as sum of layers that are pairwise mutually comonotonic risks.

Let $X \in \Gamma$ be a discrete risk with finitely many mass points: then, there exist a positive integer m, a finite sequence $\{x_j\}$, $(j = 0, \cdots, m)$, $0 \equiv x_0 < x_1 < \ldots < x_m \equiv c$ and a finite sequence $\{p_j\}$, $(j = 0, \cdots, m-1)$, $1 \geq p_0 > p_1 > p_2 > \ldots > p_{m-1} > 0$ such that the tail function H_X of X is so defined

$$H_X(x) = \sum_{j=0}^{m-1} p_j \, I_{(x_j \leq x < x_{j+1})}, \quad x \geq 0, \tag{6}$$

where $I_{(x_j \leq x < x_{j+1})}$ is the indicator function of the set $\{x : x_j \leq x < x_{j+1}\}$. Then

$$X = \sum_{j=0}^{m-1} L(x_j, x_{j+1}), \tag{7}$$

where a layer at (x_j, x_{j+1}) of X is defined as the loss from an excess-of-loss cover, namely

$$L(x_j, x_{j+1}) = \begin{cases} 0 & 0 \leq X \leq x_j \\ X - x_j & x_j < X < x_{j+1} \\ x_{j+1} - x_j & X \geq x_{j+1} \end{cases} \tag{8}$$

and the tail function of the layer $L(x_j, x_{j+1})$ is given by

$$H_{L(x_j, x_{j+1})}(x) = \begin{cases} p_j & 0 \leq x < x_{j+1} - x_j \\ 0 & x \geq x_{j+1} - x_j \end{cases}. \tag{9}$$

If we consider a Bernoulli random variable B_{p_j} such that $Pr[B_{p_j} = 1] = p_j = 1 - Pr[B_{p_j} = 0]$ then $L(x_j, x_{j+1})$ is a two-points distributed random variable which satisfies the equality in distribution $L(x_j, x_{j+1}) \stackrel{\mathrm{d}}{=} (x_j - x_{j+1}) \, B_{p_j}$.

Additivity for comonotonic risks and positive homogeneity of distorted risk measures W_g ensure that

$$W_g(X) = \sum_{j=0}^{m-1} W_g(L(x_j, x_{j+1})) = \sum_{j=0}^{m-1} (x_{j+1} - x_j) \, g(p_j).$$

In this way for any discrete risk $X \in \Gamma$ for which representation (6) holds and any distortion function g, we can assert that

$$W_g(X) = \sum_{j=0}^{m-1} (x_{j+1} - x_j) \, g(p_j). \tag{10}$$

3 The Class of Concave Distortion Risk Measures

In the particular case of a concave distortion measure, the related distortion risk measure satisfying properties *P1-P4* is also sub-additive and it preserves stop-loss order. As it is well known, examples of concave distortion risk measures are the *Tail Value-at-Risk* and the *PH-transform risk measure*, whereas *quantile risk measure* is not a concave risk measure.

In the previous section we deduced an explicit formula for the distortion risk measure $W_g(X)$ when a discrete risk $X \in \Gamma$ with finitely many mass points is considered. This result may be used to obtain upper and lower bounds for sums of discrete and identically distributed risks with common tail function given by (6) when we consider the following framework where to set the study: the Fréchet space consisting of all n-dimensional random vectors $\mathbf{X}$ possessing $(H_{X_1}, H_{X_2}, \ldots, H_{X_n})$ as marginal tail functions, for which the condition $\sum_{i=1}^{n} H_{X_i}(0) \leq 1$ is fulfilled and the distortion function g is assumed to be concave.

3.1 Upper bound for sums of discrete and identically distributed risks

Let $\mathbf{X}$ be a random vector with discrete and identically distributed risks $X_i \in \Gamma$. The least attractive random vector with given marginal distribution functions has the comonotonic joint distribution (see e.g. Dhaene et. al. [DDGKV02a] and Kaas et al. [KDG00]), namely

$$W_g(S) \leq W_g(S^c).$$

Now we want to give an explicit formula for $W_g(S^c)$. Let the common tail function of X_i be written as

$$H_{X_i}(x) = \sum_{j=0}^{m-1} p_j \, I_{(x_j \leq x < x_{j+1})}, \quad x \geq 0, \tag{11}$$

where m is a positive integer and $1 \geq p_0 > p_1 > p_2 > \ldots > p_{m-1} > 0$, $0 \equiv x_0 < x_1 < \ldots < x_m \equiv c$. Then

$$S^c \stackrel{\mathrm{d}}{=} n\, X_1,$$

and by subadditivity of the concave risk measure W_g it follows that

$$W_g(S) \leq W_g(S^c) = \sum_{j=0}^{m-1} (x_{j+1} - x_j)\, n\, g(p_j).$$

Namely, under representation (11) the riskiest portfolio S^c exhibits the following risk measure

$$\sum_{j=0}^{m-1} (x_{j+1} - x_j)\, n\, g(p_j). \tag{12}$$

3.2 Lower bound for sums of discrete and identically distributed risks

As in the previous subsection, let $\mathbf{X}$ be a random vector with discrete and identically distributed risks $X_i \in \Gamma$. In the Fréchet space consisting of all n-dimensional random vectors $\mathbf{X}$ possessing $(H_{X_1}, H_{X_2}, \ldots, H_{X_n})$ as marginal tail functions, for which the condition $\sum_{i=1}^{n} H_{X_i}(0) \leq 1$ is fulfilled, the safest random vector is given by (see Dhaene et Denuit, [DD99]) the vector $\mathbf{X}^e = (X_1^e, X_2^e, \ldots, X_n^e)$ where the components are said to be mutually exclusive because $Pr[X_i^e > 0, X_j^e > 0] = 0$ for all $i \neq j$. Let S^e denote the sum of mutually exclusive risks $X_1^e, X_2^e, \ldots, X_n^e$. In order to have an explicit formula for $W_g(S^e)$, note that its tail function is given by

$$H_{S^e}(x) = \sum_{i=1}^{n} H_{X_i}(x), \quad \text{for all } x \geq 0. \tag{13}$$

Owing to the fact that the common tail function of X_i is written as (11) where $np_0 \leq 1$, the tail function of the sum S^e of mutually exclusive risks becomes

$$H_{S^e}(x) = n \sum_{j=0}^{m-1} p_j \, I_{(x_j \leq x < x_{j+1})}, \quad \text{for all } x \geq 0. \tag{14}$$

Note that S^e can be written as a sum of layers

$$S^e = \sum_{j=0}^{m-1} \tilde{L}(x_j, x_{j+1}),$$

where $\tilde{L}(x_j, x_{j+1})$ is a two-points distribution with $\tilde{L}(x_j, x_{j+1}) \stackrel{d}{=} B_{n\,p_j}$. By considering a concave distortion risk measure, it is

$$W_g(S) \geq W_g(S^e) = \sum_{j=0}^{m-1} (x_{j+1} - x_j)\, g(n\, p_j). \tag{15}$$

In other words, under hypothesis (11) where $np_0 \leq 1$, the safest portfolio S^e exhibits the following risk measure

$$\sum_{j=0}^{m-1} (x_{j+1} - x_j)\, g(n\, p_j). \tag{16}$$

4 Optimal Gap Between Bounds of Risk Measures

In the previous section lower and upper bounds for sums of discrete and identically distributed risks $X_i \in \Gamma$ have been obtained. Attention now is devoted to the study of the difference between risk measures of upper and lower bounds, with the aim of obtaining some information on random variables for which the gap is maximum or minimum. Starting from formulations (16) and (12) exhibiting bounds for aggregate claims S of the portfolio $\mathbf{X}$, we face the problem of studying the difference between bounds in order to minimize and maximize it. The problem in p_j $(j = 1, \cdots, m-1)$ and x_j $(j = 1, \cdots, m)$ is then

$$\max / \min \sum_{j=0}^{m-1} (x_{j+1} - x_j)[ng(p_j) - g(np_j)], \tag{17}$$

where

$x_0 \equiv 0;\ x_j < x_{j+1};\ x_m \equiv c;$
$0 < p_{m-1} < \cdots < p_0 \leq \frac{1}{n};\ n > 1;$
$g : [0, 1] \to [0, 1];\ g(0) = 0;\ g(1) = 1;$ g is non-decreasing and concave.

Note that the optimization problem is related to the maximization/minimization of the gap between upper and lower bounds for risk measures, namely to the maximization/minimization of the difference $W_g(S^c) - W_g(S^e)$.

Let $\phi_n(p_j)$ be equal to $ng(p_j) - g(np_j)$: since g is concave and $g(0) = 0$, ϕ_n is non-negative; in particular $\phi_n(0) = 0$. Moreover, non-negativity and concavity of g imply that ϕ_n is non-decreasing. After setting $\Delta_j = x_{j+1} - x_j$ the problem becomes

$$\max / \min \sum_{j=0}^{m-1} \Delta_j\ \phi_n(p_j), \tag{18}$$

where

$\Delta_j > 0;\ \sum_{j=0}^{m-1} \Delta_j = c;$
$0 < p_{m-1} < \cdots < p_0 \leq \frac{1}{n};\ n > 1;$
$\phi_n : [0, \frac{1}{n}] \to \mathcal{R};\ \phi_n(0) = 0;\ \phi_n(\frac{1}{n}) = ng(\frac{1}{n}) - 1 \geq 0;$ ϕ_n is non-decreasing.

Note that the feasible set is not closed, so at any first step a relaxed problem with closed constraints will be faced.

4.1 The problem of minimizing the gap

The problem of minimizing the difference $W_g(S^c) - W_g(S^e)$ may be faced both in terms of p_j both in terms of Δ_j.

a) *Solution with respect to the* p_j*s.*
   At a first step the hypothesis that the constraints for the p_js admit equality is assumed (namely we consider the closure of the feasible set); by monotonicity of ϕ_n it follows that $p_j = 0$; moreover, if there exists $0 < \epsilon < \frac{1}{n}$ such that $\phi_n(\epsilon) = 0$ then all the p_js could be set in the interval $(0, \epsilon]$ and a solution would exist also in the original open set.

b) *Solution with respect to the* Δ_j*s.*
Starting from the case that the constraints for the Δ_js admit equality (the closure of the feasible set is assumed), by non-decreasing monotonicity of ϕ_n it follows that the minimum is when $\Delta_{m-1} = c$ and all the other Δ_j are set equal to 0 (that is $x_0 = x_1 = \cdots = x_{m-1} = 0$ and $x_m = c$). In the particular case of a constant function ϕ_n in the interval $[p_{m-1}, p_0]$, the problem would admit interior minima given by any feasible choice of the Δ_js.

4.2 The problem of maximizing the gap

The problem of maximizing the difference $W_g(S^c) - W_g(S^e)$ exhibits the following solutions in terms of p_j and in terms of Δ_j.

a) *Solution with respect to the* p_j*s.*
By referring to the case of closed feasible set (that is the constraints for the p_js admit equality) the optimal solution is given by $p_j = \frac{1}{n}$; if, moreover, there exists $0 < \epsilon < \frac{1}{n}$ such that ϕ_n is constant in the interval $[\frac{1}{n} - \epsilon, \frac{1}{n}]$, then all the p_js could be set in that interval and a solution would exist also in the original open set.

b) *Solution with respect to the* Δ_j*s.*
Under the relaxed hypothesis of equality constraints on Δ_js, the maximum is when $\Delta_0 = c$ and all the other Δ_js are set equal to 0 (that is $x_0 = 0$ and $x_1 = \cdots = x_m = c$). Notc that if ϕ_n wcrc constant in thc intcrval $[p_{m-1}, p_0]$ the problem would admit interior maxima: any feasible choice of the Δ_js is solution.

5 Concluding Remarks

In this paper we face the problem of studying the gap between bounds for risk measures of sums of discrete and identically distributed risks. Starting from the representation of risks as sums of layers, explicit formulas for risk measures of upper and lower bounds of sums of risks are obtained in the particular case of concave distortion risk measures. A maximization/minimization problem related to the maximization/minimization of the gap between risk measures of upper and lower bounds is solved with respect to information characterizing the random vector **X**.

References

[CF05] Campana, A., Ferretti, P.: On distortion risk measures for sums of discrete and identically distributed risks. Giornale dell'Istituto Italiano degli Attuari, **LXVIII**: 89–104 (2005)

[DD99] Dhaene, J., Denuit, M.: The safest dependence structure among risks. Insurance: Mathematics and Economics, **25**: 11–21 (1999)

[DDGKV02a] Dhaene, J., Denuit, M., Goovaerts, M.J., Kaas, R., Vynckce, D.: The concept of comonotonicity in actuarial science and finance: theory. Insurance: Mathematics and Economics, **31**: 3–33 (2002)
[DDGKV02b] Dhaene, J., Denuit, M., Goovaerts, M. J., Kaas, R., Vynckce, D.: The concept of comonotonicity in actuarial science and finance: applications. Insurance: Mathematics and Economics, **31**: 133–161 (2002)
[DVTGKV04] Dhaene, J., Vanduffel, S., Tang, Q. H., Goovaerts, M. J., Kaas, R., Vyncke, D.: Solvency capital, risk measures and comonotonicity: a review. Research Report OR 0416, Department of Applied Economics, K.U. Leuven (2004)
[KDG00] Kaas, R., Dhaene, J., Goovaerts, M. J.: Upper and lower bounds for sums of random variables. Insurance: Mathematics and Economics, **27**: 151–168 (2000)
[W95] Wang, S.: Insurance pricing and increased limits ratemaking by proportional hazard transforms. Insurance: Mathematics and Economics, **17**: 43–54 (1995)
[W96] Wang, S.: Premium calculation by transforming the layer premium density. ASTIN Bulletin, **26**: 71–92 (1996)

Characterization of Convex Premium Principles

Marta Cardin and Graziella Pacelli

Summary. In actuarial literature the properties of risk measures or insurance premium principles have been extensively studied. We propose a characterization of a particular class of coherent risk measures defined in [Art99]. The considered premium principles are obtained by expansion of TVaR measures, consequently they look very interesting in insurance pricing where TVaR measures are frequently used to value tail risks.

Key words: Risk measures; Premium principles; Choquet measures; Distortion function; TVaR.

1 Introduction

Premium principles are the most important risk measures in actuarial sciences and frequently the insurers are also interested in measuring the upper tails of distribution functions [Wan97]. There are different methods that actuaries use to develop premium principles [Den05].

In this paper we propose an axiomatic approach based on a minimal set of properties which characterizes an insurance premium principle as a Choquet integral with respect to a distorted probability. As it is well known, distortion risk measures are introduced in the actuarial literature by Wang [Wan96] and are related to the coherent risk measures. Two particular examples of Wang risk measures are given by V_α and $TVaR_\alpha$, (Value at Risk at level α and Tail Value at Risk at level α respectively). The distortion function giving rise to the V_α is not concave, so that V_α is not a coherent measure, while the distortion function giving rise to the $TVaR_\alpha$, is concave so that $TVaR_\alpha$ is a coherent measure $TVaR_\alpha$ [Den05]. The importance in actuarial science and in finance of $TVaR_\alpha$, as a measure of the upper tail of a distribution function is well known and we refer to [Den05].

In this paper we consider a rather general set of risks and for the premium principles we ask some natural assumptions, **(A1)** – **(A4)**. We obtain for all the premium principles of this class, an integral representation by a non-additive convex measure and then an integral representation by concave distortion functions so that the con-

Cira Perna, Marilena Sibillo (Eds.): Mathematical and Statistical Methods in Insurance and Finance

sidered premium principles are a convex combination of coherent risk measures as $TVaR_\alpha, \alpha \in [0, 1]$.

The paper is organized as follows. In Section 2 we provide some necessary preliminaries and we introduce the properties that characterize the premium principles considered in this paper. In Section 3 we recall some basic facts of Choquet expected utility and we introduce a modified version of Greco Theorem [Gre82]. In Section 4 we present distortion risk measures. Finally in Section 5 we obtain the integral representation result premium principles and the characterization as convex combination of $TVaR_\alpha, \alpha \in [0, 1]$.

2 Insurance Premium Principles

In actuarial applications a risk is represented by a non-negative random variable. We consider an insurance contract in a specified time period $[0, T]$. Let Ω be the state space and $\mathcal{F}$ the event σ-field at the time T. Let $\mathbf{P}$ be a probability measure on $\mathcal{F}$. We consider an insurance contract described by a non-negative random variable X, $X : \Omega \to \mathbb{R}$ where $X(\omega)$ represents its payoff at time T if state ω occurs. We denote by F_X the distribution function of X and by S_X the survival function.

Frequently an insurance contract provides a franchise, and it is then interesting to consider the values ω such that $X(\omega) > a$: in this case the contract pays for $X(\omega) > a$ and nothing else. It is then useful to consider the random variable

$$(X - a)_+ = \max(X(\omega) - a, 0). \tag{1}$$

Consider a set, L of non-negative random variables with the following property:

i) $\quad aX, \quad (X - a)_+, \quad (X - (X - a)_+) \in L \qquad \forall X \in L$, and $a \in [0, +\infty)$.

We observe that such a set L is not necessarily a vector space.

We denote the insurance prices of the contracts of L by a functional H, where

$$H : L \to \widetilde{\mathbb{R}} \tag{2}$$

and $\widetilde{\mathbb{R}}$ is the extended real line. We consider some properties that it is reasonable to assume for an insurance functional price H:

(P1) $H(X) \geq 0$ for all $X \in L$.

(P2) If $c \in [0, +\infty)$ then $H(c) = c$.

(P3) $H(X) \leq \sup_{\omega\in\Omega} X(\omega)$ for all $X \in L$.

(P4) $H(aX + b) = aH(X) + b$ for all $X \in L$ such that $aX + b \in L$ with $a, b \in [0, +\infty)$.

(P5) If $X(\omega) \le Y(\omega)$ for all $\omega \in \Omega$ for $X, Y \in L$ then $H(X) \le H(Y)$.

(P6) $H(X + Y) \le H(X) + H(Y)$ for all $X, Y \in L$ such that $X + Y \in L$.

We observe that the properties **(P4)** and **(P6)** imply the following property:

(P7) $H(aX + (1 - a)Y) \le aH(X) + (1 - a)H(Y)$ for all $X, Y \in L$ and $a \in [0, 1]$ such that $aX + (1 - a)Y \in L$.

The last property is the convexity property and it means that diversification does not increase the total risk. In the insurance context this property allows the pooling of risk effects.

Now we present some assumptions which are frequently in the applications.

(A1) $H(X) = H(X - (X - a)_+) + H((X - a)_+)$ for all $X \in L$ and $a \in [0, +\infty)$.

This condition splits a risk X into two comonotonic parts (see for example [Den94]), and permits the identification of the part of the premium charged for the risk with the reinsurance premium charged by the reinsurer.

(A2) If $E(X - a)_+ \le E(Y - a)_+$ for all $a \in [0, +\infty)$ then $H(X) \le H(Y)$ for all $X, Y \in L$.

In other words the functional price H respects the stop-loss order. Remember that stop-loss order considers the weight in the tail of distributions; when other characteristics are equals, stop-loss order selects the risk with less heavy tails [Wir99].

(A3) The price, $H(X)$, of the insurance contract X depends only on its distribution F_X.

Frequently, this hypothesis is assumed in literature, see for example [Wan97] and [Kus01]. The hypothesis **(A3)** states that it is not the particular scenario that determines the price of a risk, but the probability distribution of X assigns the price to X. So risks with identical distributions have the same price.

Finally, we present a continuity property that is used in characterizing certain premium principles.

(A4) $\lim_{n \to +\infty} H(X - (X - n)_+) = H(X)$ for all $X, Y \in L$.

3 Choquet Pricing of Insurance Risks

The development of premium functionals based on the Choquet integration theory has gained considerable interest in recent years when there is ambiguity on the loss distribution or when there is correlation between the individual risks in the case. In fact the traditional pricing functionals may be inadequate to determine the premiums that cover the risk.

Capacities are set functions defined on 2^Ω to real values which generalize the notion of probability distribution. Formally, a capacity is a normalized monotone function, for the definition and properties see for example [Den94] and [Den05].

As is well known, the Choquet integral has been extensively applied in the context of decision under uncertainty and in risk applications.

Definition 1. *Let υ be a capacity $\upsilon : 2^\Omega \to \mathbb{R}^+$ and X a random non-negative variable defined on $(\Omega, \mathcal{F})$, then the Choquet integral of X respect to υ is defined as*

$$\int_\Omega X d\upsilon = \int_0^{+\infty} \upsilon\{\omega : X(\omega) > x\} dx. \tag{3}$$

We now give the representation theorem for the functional H which satisfies some properties of the list above. This result is a new version of the well known Greco theorem (see [Gre82]), and our new assumptions match perfectly with an actuarial point of view.

Theorem 1. *Let L be a set of non-negative random variables such that L has property i) and we suppose that $I_\Omega \in L$ where I_Ω is the indicator function of Ω. We consider a premium principle $H : L \to \widetilde{\mathbb{R}}$ such that:*

a) $H(I_\Omega) < +\infty$,
b) H satisfies the hypotheses (**P5**), (**A1**) *and* (**A4**).

Then there exists a capacity $\upsilon : 2^\Omega \to \mathbb{R}^+$ such that for all $X \in L$

$$H(X) = \int_\Omega X d\upsilon = \int_0^{+\infty} \upsilon\{\omega : X(\omega) > x\} dx. \tag{4}$$

Proof: By the Representation Theorem and Proposition 1.2 of [Gre82] there exists a capacity υ such that $H(X) = \int_\Omega X d\upsilon$.

The following result considers the convex capacities.

Corollary 1. *Let L be a set of non-negative random variables such that L has property i). Let H be a premium principle that satisfies the hypotheses of Theorem 1, and verifies the property* (**P6**)*, then there exists a convex capacity in* (4).

Proof: The thesis follows, in fact, that it is well known that v is convex if and only if H is subadditive [Wan98].

Remark. We observe that:

i) From Theorem 1 follows the property (**P1**) and (**P3**) for H.
ii) The property (**P4**) for $b = 0$ is also according to Theorem 1. From Theorem 1 we have property (**P2**), then (**P4**) follows for every b, and then (**P7**) follows as well.

4 Distortion Risk Measures

In this section we report some well known risk measures and present the distortion functions measure for some of them. Distortion premium principles have been extensively studied in the past several years, see for example [Wan98], [Wir99] and [Wu03].

If X is a random variable, the quantile reserve at 100αth percentile or Value at Risk is

$$V_\alpha(X) = \inf\{x \in \mathbb{R} \mid F_X(x) \geq \alpha\} \qquad \alpha \in (0, 1). \tag{5}$$

A single quantile risk measure of a fixed level α does not give information about the thickness of the upper tail of the distribution function of X, so that other measures are considered.
In particular we consider the Tail Value at Risk at level α, $TVaR_\alpha(X)$, is defined as:

$$TVaR_\alpha(X) = \frac{1}{(1-\alpha)} \int_\alpha^1 V_\alpha(X) d\alpha \qquad \alpha \in (0, 1). \tag{6}$$

It is known, that given a non-negative random variable X, for any increasing function f, with $f(0) = 0$ and $f(1) = 1$, we can define a premium principle

$$H(X) = \int_0^{+\infty} (1 - f(F_X(t))dt = \int_0^{+\infty} g(S_X(t))dt = \int_\Omega X dv, \tag{7}$$

where g is a distortion function, $g(x)$= $1 - f(1-x)$ and $v = go\mathbf{P}$.

Remark. All distortion premium principles have the properties (**P1**), (**P2**), (**P3**) and (**P4**). If g is concave (f convex), then H also satisfies the property (**P6**) and (**P7**) follows.

It is well known that the quantile Value at Risk, (5), is a distorted risk measure, while TailVar is a convex distorted risk measure. In fact, $TVaR_\alpha$ can be obtained by (7) where f is the function defined as follows:

$$f(u) = \begin{cases} 0 & u < \alpha, \\ \frac{(u-\alpha)}{(1-\alpha)} & u \geq \alpha. \end{cases} \tag{8}$$

5 Representation of a Class of Premium Functionals

Now we provide a characterization of the class of continuous increasing and convex functions.

Proposition 1. *If f is a continuous increasing convex function, defined on* $[0, 1]$ *then there exists a probability measure* μ *on* $[0, 1]$ *such that*

$$f(x) = \int_0^1 \frac{(x-\alpha)_+}{(1-\alpha)} d\mu(\alpha) \tag{9}$$

for $\alpha \in [0, 1]$.

Proof: Given f, a continuous increasing convex function with $f(0) = 0$, there then exists a non-negative measure ν on $[0, 1]$ such that

$$f(x) = \int_0^x (x-\alpha) d\nu(\alpha). \tag{10}$$

We can write

$$f(x) = \int_0^1 (x-\alpha)_+ d\nu(\alpha). \tag{11}$$

Then, a probability measure μ on $[0, 1]$ exists, such that

$$f(x) = \int_0^1 \frac{(x-\alpha)_+}{(1-\alpha)} d\mu(\alpha), \qquad \alpha \in [0, 1]. \tag{12}$$

Theorem 2. *Let L be a set of non-negative random variables such that L has property i). We suppose that $I_\Omega \in L$ where I_Ω is the indicator function of Ω and $I_{X \geq a} \in L$ for any $X \in L$ and any $a > 0$. We consider a premium principle $H : L \to \widetilde{\mathbb{R}}$ such that:*

a) $H(I_\Omega) < +\infty$,
b) *H satisfies the hypotheses* **(A1)** – **(A4)**.

Then there exists a probability measure m on $[0, 1]$ *such that:*

$$H(X) = \int_0^1 TVaR_\alpha(X) dm(\alpha). \tag{13}$$

Proof: From Theorem 1 we can conclude that there exists a capacity, such that for all $X \in L$,

$$H(X) = \int_\Omega X dv. \tag{14}$$

Since H is the comonotonic additive (see [Den94]), and H verifies **(A2)**, then H is sub-additive, i.e. H has the property **(P6)** ([Wan98]). It follows from Corollary 1, that the capacity v in (14) is convex. Since $I_{X>a} \in L$ for any $X \in L$ and any $a > 0$, from Corollary 3.1 of [Wu03] it follows that a convex increasing function exists, $f : [0, 1] \to [0, 1]$ with $f(0) = 0$ and $f(1) = 1$ i.e. f such that

$$H(X) = \int_0^{+\infty} (1 - f(F_X(t))dt. \tag{15}$$

From Proposition 1, it follows that a probability measure $m(\alpha)$ exists such that f can be represented

$$f(x) = \int_0^1 \frac{(x-\alpha)_+}{(1-\alpha)} dm(\alpha) \tag{16}$$

for $\alpha \in [0, 1]$, and $f(1) = 1$.

Then, interchanging the integrals for the Fubini Theorem for every $X \in L$:

$$\begin{aligned} H(X) &= \int_0^{+\infty} (1 - f(F_X(t))dt = \int_0^{+\infty} [1 - \int_0^1 \frac{(F_X(t)-\alpha)_+}{(1-\alpha)} dm(\alpha)]dt = \\ &= \int_0^{+\infty} dt \int_0^1 [1 - \frac{(F_X(t)-\alpha)_+}{(1-\alpha)}]dm(\alpha) = \\ &= \int_0^1 dm(\alpha) \int_0^{+\infty} dt[1 - \frac{(F_X(t)-\alpha)_+}{(1-\alpha)}] = \\ &= \int_0^1 dm(\alpha) TVaR_\alpha. \end{aligned} \tag{17}$$

The results obtained for the class of insurance functional prices seem interesting, both because the class of functionals is determined from few natural properties and these functional prices follow closely linked together to a well known risk measure as $TVaR_\alpha, \alpha \in [0, 1]$. Moreover, we point out that the most important properties for a functional price follow easily from the obtained representation.

References

[Art99] Artzner, P., Delbaen, F., Eber, J. M., Heath, D.: Coherent measures of risks, Mathematical Finance, **9**, 203–228 (1999)

[Den94] Denneberg, D.: Non-additive measure and integral Kluwer Academic Publishers (1994)

[Den05] Denuit, M., Dhaene, J., Goovaerts, M. J., Kaas, R.: Actuarial theory for Dependent Risks John Wiley & Sons Ltd. (2005)

[Gre82] Greco, G. H.: Sulla rappresentazione di funzionali mediante integrali: Rendiconti Seminario Matematico Università Padova, **66**, 21–42 (1982)

[Kus01] Kusuoka, S.: On law invariant risk measures: Advances in Mathematical Economics, **3**, 83–95 (2001)

[Wan96] Wang, S.: Premium Calculation by Transforming the Layer Premium Density, Astin Bulletin, **26**, 71–92 (1996)

[Wan98] Wang, S., Dhaene, D.: Comonotonicity correlation order and premium principle, Insurance: Mathematics and Economics, **22**, 235–242 (1998)

[Wan97] Wang, S., Young, V. R., Panjer, H. H.: Axiomatic characterization of insurance prices, Insurance: Mathematics and Economics, **21**, 173–183 (1998)

[Wir99] Wirch, L. J., Hardy, M. R.: A synthesis of risk measures for capital adequacy, Insurance: Mathematics and Economics, **25**, 337–347 (1999)

[Wu03] Wu, X., Wang, J.: On Characterization of distortion premium, Astin Bulletin, **33**, 1–10 (2003)

FFT, Extreme Value Theory and Simulation to Model Non-Life Insurance Claims Dependences

Rocco Roberto Cerchiara

Summary. This paper shows an example of integrated use of three different approaches (Extreme Value Theory (EVT), two-dimensional Fast Fourier Transform (FFT) and Monte Carlo simulation) to model non-life insurance company aggregate losses, taking into account the need for Internal Risk Model development in the light of Solvency II European project. In particular EVT permits the definition of the truncation point between small and large claims. Two-dimensional FFT is used to model not only aggregate losses, but dependence between its basic components too. Finally, Monte Carlo simulation describes large claims behaviour. Collective Risk Model has been developed using Matlab software.

Key words: Non-life insurance; Aggregate losses; Risk theory; Two-dimensional fast Fourier transform; Extreme value theory; Simulation.

1 Introduction

Homer and Clark (see [HC04]) has shown a case study for a non-life insurance company which has developed the solvency assessment model, based on simulation and the Collective Risk Theory approach, with the goal of calculating the aggregate losses probability distribution. The model made the assumption that small and large aggregate losses are independent. Very small and large losses were simulated separately and the results were combined. Typically this assumption is not true. This paper shows an example, based on [HC04] results, of the integrated use of three different approaches, where EVT permits the definition of the truncation point between small and large claims. Two-dimensional FFT allows modelling not only of aggregate losses, but dependence between its basic components too, and Monte Carlo simulation describes large claims behaviour. The model works in an efficient way preserving dependence structure. In the next section, total aggregate loss expected value produced from small and large claims, $E[X^{Small,Large}]$, will be calculated using both the example shown in [HC04] and the Danish Fire Insurance Database, used in [Cer06], in relation to the period 1980-2002, where amounts are expressed in DK[1]. Calculation algorithms have been developed using Matlab.

[1] The Danish Fire Insurance Database was obtained from Mette Rytgaard

Cira Perna, Marilena Sibillo (Eds.): Mathematical and Statistical Methods in Insurance and Finance

2 An Example of EVT, FFT and Simulation Application

In Collective Risk Model (see [DPP94] and [KPW98]), the aggregate loss random variable X in one year is:

$$X = \sum_{i=1}^{k} Z_i, \tag{1}$$

where typically, k is a claim number random variable, independent of claim size random variables Z_i (i.i.d.).

Using the Danish Fire Database and EVT, a possible threshold is 10×10^6 (see [Cer06] and also [EKM97], where only 1980–1993 has been considered). This truncation point separates small and large claims.

Consider a claim size distribution Z with 6 points ($0, 2 \times 10^6, 4 \times 10^6, 6 \times 10^6, 8 \times 10^6, 10 \times 10^6$) and the following probabilities: 0.0%, 43.8%, 24.6%, 13.8%, 7.8%, 10.0% (note that FFT is applicable for discrete random variable Z and X). It is important to remember that this paper is a brief example, but for the real data considered, Z would need to be expanded to a greater number of points. This probability distribution can be represented using the bivariate matrix M_Z (see [HC04] for more details), which permits the application of two-dimensional FFT. In particular, the first column of Table 1 shows Z^{Small} probability distribution with exclusion of the truncation point 10×10^6, defined before with EVT. In cell (0, 1) there is the probability that claim size is above 10×10^6 (10%).

This representation gives an important example of how two-dimensional FFT is flexible, because it can work on variables measured in different ways (number and amount). If there is dependence between small and large claims, and one claim frequency is used to generate total claim number, then covariance of both groups aggregate loss random variables, X^{Small} and X^{Large} (see [HC04]), is:

$$Cov(X^{Small}, X^{Large}) = p \cdot E\left[Z^{Small}\right] \times (1-p) \cdot E\left[Z^{Large}\right] \cdot \left(\sigma_k^2 - E[k]\right) \tag{2}$$

where p is the probability of having a small claim (see [Wan98] for an interesting comparation of Copule and FFT to model dependences). If claim number k follows Binomial Negative distribution, the previous formula is positive; if k has Poisson distribution then covariance is zero. In this paper, under dependence assumption, Binomial Negative distribution is used for k with mean = 10 and variance = 20 (hypothesis; while for the real data considered, the overall frequency is much higher, it will, however, be continued with this simplified assumption for clarity).

Table 1. Bivariate Matrix M_Z of small claim size Z^{Small} and large claim number k^{Large}.

	k^{Large}					
Z^{Small}	0	1	2	3	4	5
0,000,000	0.000	0.100	0.000	0.000	0.000	0.000
2,000,000	0.438	0.000	0.000	0.000	0.000	0.000
4,000,000	0.246	0.000	0.000	0.000	0.000	0.000
6,000,000	0.138	0.000	0.000	0.000	0.000	0.000
8.000,000	0.078	0.000	0.000	0.000	0.000	0.000
10,000,000	0.000	0.000	0.000	0.000	0.000	0.000

Table 2. Bivariate Distribution of X^{Small} and k^{Large}.

	k^{Large} (%)							$Pr(X^{Small})$
X^{Small}	0	1	2	3	4	5	6	%
0,000,000	0.37	0.15	0.03	0.01	0.00	0.00	0.00	0.6
2,000,000	0.57	0.25	0.07	0.02	0.00	0.00	0.00	0.9
4,000,000	0.74	0.36	0.09	0.03	0.00	0.00	0.00	1.2
6,000,000	0.91	0.50	0.14	0.04	0.00	0.00	0.00	1.6
8,000,000	1.19	0.65	0.20	0.06	0.01	0.00	0.00	2.1
10,000,000	1.44	0.83	0.28	0.08	0.01	0.00	0.00	2.6
12,000,000	1.71	1.03	0.36	0.10	0.02	0.00	0.00	3.2
14,000,000	1.87	1.23	0.45	0.13	0.03	0.00	0.00	3.7
16,000,000	2.10	1.42	0.54	0.16	0.03	0.01	0.00	4.3
18,000,000	2.23	1.78	0.62	0.19	0.04	0.01	0.00	4.9
20,000,000	2.41	1.82	0.70	0.21	0.05	0.01	0.00	5.2
22,000,000	2.46	1.83	0.77	0.24	0.06	0.01	0.00	5.4
24,000,000	2.55	1.90	0.82	0.26	0.06	0.01	0.00	5.6
26,000,000	2.44	1.94	0.86	0.28	0.07	0.02	0.00	5.6
28,000,000	2.39	1.95	0.88	0.30	0.08	0.02	0.00	5.6
30,000,000	2.31	1.93	0.89	0.31	0.08	0.02	0.00	5.5
32,000,000	2.21	1.87	0.89	0.31	0.08	0.02	0.00	5.4
34,000,000	2.09	1.80	0.87	0.31	0.09	0.02	0.00	5.2
36,000,000	1.95	1.71	0.84	0.31	0.09	0.02	0.00	4.9
38,000,000	1.81	1.61	0.81	0.30	0.09	0.02	0.01	4.6
40,000,000	1.70	1.7	0.76	0.29	0.08	0.02	0.01	4.4
42,000,000	1.66	1.55	0.72	0.28	0.08	0.02	0.01	4.3
44,000,000	1.40	1.26	0.66	0.26	0.08	0.02	0.00	3.7
46,000,000	1.32	1.20	0.61	0.25	0.08	0.02	0.00	3.5
48,000,000	1.16	1.04	0.56	0.23	0.07	0.02	0.00	3.1
50,000,000	1.03	0.93	0.50	0.21	0.07	0.02	0.00	2.8
							Total	100.0

Bivariate aggregate distribution will be:

$$M_X = IFFT(PGF(FFT(M_Z))) \ and \ \ PGF = (2-t)^{-10} \qquad (3)$$

where PGF is Probability Generation Function of Binomial Negative distribution, FFT is two-dimensional Fast Fourier Transform procedure and IFFT is two-dimensional Inverse FFT (see [HC04], [KPW98] and [Wan98]).

Matrix M_X (see Table 2) gives bivariate distribution of large claim number, k^{Large}, and small claim aggregate loss, X^{Small} (discretized with 26 points). Marginal distribution of X^{Small} is obtained as the sum of probabilities of each row.

Table 3. Rescaling of Table 2.

	k^{Large} (%)							Row Total
X^{Small}	0	1	2	3	4	5	6	%
0,000,000	64.7	26.2	5.2	1.6	0.8	0.8	0.8	100.00
2,000,000	62.4	27.4	7.2	1.6	0.4	0.4	0.4	100.00
4,000,000	60.3	29.3	7.3	2.0	0.3	0.3	0.3	100.00
6,000,000	57.0	31.3	8.8	2.2	0.2	0.2	0.2	100.00
8,000,000	56.3	30.8	9.5	2.6	0.5	0.2	0.2	100.00
10,000,000	54.5	31.4	10.6	2.8	0.4	0.1	0.1	100.00
12,000,000	53.1	32.0	11.2	2.9	0.6	0.1	0.1	100.00
14,000,000	50.4	33.1	12.1	3.4	0.8	0.1	0.1	100.00
16,000,000	49.3	33.3	12.7	3.6	0.7	0.3	0.1	100.00
18,000,000	45.8	36.5	12.7	3.8	0.8	0.3	0.1	100.00
20,000,000	46.3	35.0	13.5	3.9	1.0	0.3	0.1	100.00
22,000,000	45.8	34.1	14.3	4.4	1.1	0.3	0.1	100.00
24,000,000	45.5	33.9	14.6	4.6	1.1	0.3	0.1	100.00
26,000,000	43.5	34.6	15.3	4.9	1.2	0.3	0.1	100.00
28,000,000	42.5	34.7	15.7	5.2	1.4	0.4	0.1	100.00
30,000,000	41.7	34.8	16.1	5.5	1.4	0.4	0.1	100.00
32,000,000	41.1	34.7	16.5	5.7	1.5	0.4	0.1	100.00
34,000,000	40.3	34.7	16.8	5.9	1.8	0.4	0.1	100.00
36,000,000	39.6	34.7	17.1	6.2	1.9	0.5	0.1	100.00
38,000,000	39.0	34.7	17.4	6.4	1.9	0.5	0.1	100.00
40,000,000	38.4	35.5	17.2	6.4	1.8	0.5	0.1	100.00
42,000,000	38.5	35.9	16.7	6.4	1.9	0.5	0.1	100.00
44,000,000	38.0	34.2	17.9	6.9	2.2	0.6	0.1	100.00
46,000,000	37.9	34.5	17.5	7.1	2.2	0.6	0.1	100.00
48,000,000	37.5	33.7	18.2	7.4	2.4	0.6	0.2	100.00
50,000,000	37.3	33.8	18.1	7.4	2.5	0.7	0.2	100.00

$E[X^{Small}] = 28.0 \times 10^6$. Next, conditional frequency distribution for large claims is obtained, "rescaling" the matrix M_X, where each value is divided by the corresponding row total value.

Table 3 shows clearly dependence between small and large claims, observing that in each row there are probabilities to have $k_i = 0, 1, 2, \ldots, 6$ large claims. For example, in the column where $k_i = 0$, probabilities decrease for increasing values of X^{Small}, because if X^{Small} increases, the probability of having no large claims has to decrease.

Using the previous matrix it is possible to calculate the expected value of large claim number:

$$E[k^{Large}] = \sum_{j=1}^{26} \sum_{i=1}^{7} k_i \cdot \Pr\left\{k^{Large} = k_i / X^{Small} = x_j\right\} = 0.871. \quad (4)$$

Next, with 200,000 simulations for Z^{Large} using Generalized Pareto Distribution, with parameters $\xi = 0.5$ and $\beta = 8.8$ (see [Cer06]):

$$E[Z^{Large}] = 276.3 \times 10^6. \tag{5}$$

Finally, expected value of total aggregate loss is:

$$E[X^{Small,Large}] = E[X^{Small}] + E[k^{Large}] \cdot E[Z^{Large}] = 268.8 \times 10^6. \tag{6}$$

3 Conclusions

Solvency assessment model procedure is more efficient than the initial independence case and it is based on the following steps:

a) Defining truncation point with EVT.
b) Using two-dimensional FFT to calculate marginal distribution of X^{Small}.
c) Calculating the expected value of X^{Small} using the previous marginal distribution.
d) Calculating the expected value of large claim number $E[k^{Large}]$ using conditional distribution to X^{Small} distribution.
e) Simulating large claim size in order to obtain $E[Z^{Large}]$.

Dependence structure has also been preserved.

This example gives a simplified case study of integration between EVT, FFT and simulations for bivariate random variables which could be interesting for the Internal Risk Models definition under the Solvency II European insurance project.

References

[Cer06] Cerchiara, R. R.: Simulation Model, Fast Fourier Transform ed Extreme Value Theory to model non life insurance aggregate loss. Phd Thesis, University La Sapienza, Rome, Italy (2006) – Download on (italian language): http://padis.uniroma1.it/search.py?recid=468

[DPP94] Daykin, C. D., Pentikäinen, T., Pesonen, M.: Practical Risk Theory for Actuaries. Chapman & Hall, London (1994)

[EKM97] Embrechts, P., Kluppelbrg, C., Mikosch, T.: Modelling extremal events. Springer Verlag, Berlin (1997)

[HC04] Homer, D. L., Clark, D. R.: Insurance Applications of Bivariate Distributions. PCAS Forum, 823–852 (2004)

[KPW98] Klugman, S., Panjer, H., Willmot, G.: Loss Models - From Data to Decisions. John Wiley Sons, New York, First Edition (1998)

[Wan98] Wang, S. S.: Aggregation of correlated risk portfolios: models and algorithms. PCAS LXXXV, 848–923 (1998)

Dynamics of Financial Time Series in an Inhomogeneous Aggregation Framework

Roy Cerqueti and Giulia Rotundo

Summary. In this paper we provide a microeconomic model to investigate the long term memory of financial time series of one share. In the framework we propose, each trader selects a volume of shares to trade and a strategy. Strategies differ for the proportion of fundamentalist/chartist evaluation of price. The share price is determined by the aggregate price. The analyses of volume distribution give an insight of imitative structure among traders. The main property of this model is the functional relation between its parameters at the micro and macro level. This allows an immediate calibration of the model to the long memory degree of the time series under examination, therefore opening the way to understanding the emergence of stylized facts of the market through opinion aggregation.

Key words: Long memory; Financial time series; Fundamentalist agents; Chartist agents.

1 Introduction

Long term memory in financial data has been studied through several papers. Wide sets of analyses are available in the literature on time series of stock market indices, shares prices, price increments, volatility, returns, as well as several functions of returns (absolute returns, squared returns, powered returns). The analyses of long term memory can be refined considering different time scales as well as considering moving windows with different lenghts. The former have evidenced the multifractal structure, validating the efficient market hypothesis at long enough time scales [Be, MS and SSL]. The latter report a wide range of the self similarity parameters [BPS, BP and R]. Deviations from the gaussian case can be addressed at the microeconomic level to the lack of efficiency in markets, and to self-reinforcing imitative behaviour, as it also happens during large financial crashes due to endogenous causes [R].

We aim to provide a mathematically tractable financial market model that can give an insight into the market microstructure that captures some characteristics of financial time series. In our model, agents decide their trading price choosing from a set of price forecasts based on mixed chartist/fundamentalist strategy [FHK]. Agents will switch from one price to the other varying the volume to trade

Cira Perna, Marilena Sibillo (Eds.): Mathematical and Statistical Methods in Insurance and Finance

at the forecasted price. The distribution of the parameter that regulates the mixture chartist/fundamentalist forecasts reports the confidence of investors on each approach. We differ from [FHK] in that we don't introduce a performance index on a strategy.

Traders are not pure price takers: agents' opinions contribute to the market price in accordance with price and volume distribution. The distribution of volumes size evidences the occurrence of imitative behavior, and the spreading of the confidence of traders on "gurus". Aggregation and spreading of opinions give an insight into social interactions. Models that allow for opinion formation are mostly based on random interaction among agents, and they were refined considering constraints on social contact, as an example modeled through scale free networks. It has already been shown that the relevant number of social contacts in financial markets is very low, being between 3 and 4 [RB, AIV and VID], opening the way to lattice-based models. We will discuss at a general level the case of random interactions of agents, and then its consequences on aggregation and disaggregation of opinions on some strategies. We are not aiming at exploring the bid/ask spread: our price is just considered as the market price given by the mean of agents' prices. The proposed theoretical approach allows numerical calibration procedures to be avoided.

The paper is organized as follows. The first part of Section 2 describes the model, while Sections 2.1 and 2.2 are devoted to the study of such a model, in the case of independence and dependence, respectively. Section 3 concludes.

2 Market Price Dynamics

Consider N investors trading in the market, and assume that $\omega_{i,t}$ is the size of the order placed on the market by agent i at time t. This choice allows the modeling of individual traders as well as funds managers, that select the trading strategy on behalf of their customers. In the present analysis we consider investors who receive information from two different sources: observation of the macroeconomic fundamentals and adjustment of the forecast performed at the previous time. Other market characteristics, like the presence of a market maker, are not considered here, and they will be studied elsewhere. Let us define with $P_{i,t}$ the forecast of the market price performed by the investor i at time t. Each of them relies on a proportion of fundamentalist $P_{i,t}^f$ and of a chartist $P_{i,t}^c$ forecast. We can write

$$P_{i,t} = (1 - \beta_i)P_{i,t}^f + \beta_i P_{i,t}^c, \tag{1}$$

where β_i is sampled by a random variable β with compact support equal to $[0, 1]$, i.e. $\beta_i \sim \beta \in D[0, 1]$, for each $i = 1, \dots, N$.

Parameter β_i in (1) regulates the proportion of fundamentalist/chartist in each agent forecast. The closer β_i is to 0, the more the confidence is in the return to fundamentals. The closer β_i is to 1, the more the next price is estimated to be the actual price. The shape of the distribution used for sampling β_i gives relevant information on the overall behavior of agents.

In the fundamentalist analysis the value of the market fundamentals is known, and so the investor has complete information on the risky asset (the inverstor understands over or under estimation of price). Given the market price P_t we have the following fundamentalist forecast relation:

$$P_{i,t}^f = \nu(\tilde{P}_{i,t-1} - P_{t-1}), \tag{2}$$

where $\nu \in \mathbf{R}$ and $\tilde{P}_{i,t}$ is a series of fundamentals observed with a stochastic error from the agent i at time t, i.e.

$$\tilde{P}_{i,t} = \bar{P}_{i,t} + \alpha_{i,t},$$

with $\alpha_{i,t} = \zeta_i P_t$ and ζ_i sampled by a real random variable ζ with finite expected value $\bar{\zeta}$ and independent of β. The fundamental variables $\bar{P}_{i,t}$ can be described by the following random walk:

$$\bar{P}_{i,t} = \bar{P}_{i,t-1} + \epsilon_t, \quad \epsilon_t \sim N(0, \sigma_\epsilon^2).$$

Thus

$$P_{i,t}^f = \nu \bar{P}_{i,t-1} + \nu(\zeta_i - 1)P_{t-1}. \tag{3}$$

The chartist forecast at time t is limited to an adjustment of the forecast made by the investor at the previous time. The adjustment factor related to the i-th agent is a random variable γ_i. We assume that γ_i are i.i.d, with support in the interval $(1-\delta, 1+\delta)$, with $\delta \in [0, 1]$. Moreover, we suppose that

$$\mathbf{E}[\gamma_i] = \bar{\gamma}, \qquad i = 1, \ldots, N,$$

and γ_i are independent of ζ_i and β_i. Then we can write

$$P_{i,t}^c = \gamma_i P_{i,t-1}. \tag{4}$$

We assume that the aggregate size of the order placed by the agents at a fixed time t depends uniquely on t. We denote it as $\tilde{\omega}_t$, and we have

$$\tilde{\omega}_t = \sum_{i=1}^N \omega_{i,t}.$$

We assume that such aggregate size is uniformly bounded. Therefore, two thresholds exist, $\underline{\omega}$ and $\overline{\omega}$, such that for each $t > 0$, $\underline{\omega} < \tilde{\omega}_t < \overline{\omega}$.

Market price is given by the weighted mean of trading prices associated with the agents. The weights are given by the size of the order. We do not consider here the bid-ask spread or mechanisms related to the limit order book, these are left to future studies. In summary, we can write

$$P_t = \sum_{i=1}^N \omega_{i,t} P_{i,t}. \tag{5}$$

Then, by (1), (3) and (5)

$$P_t = \sum_{i=1}^N \omega_{i,t} \Big[\nu(1-\beta_i)\bar{P}_{i,t-1} + \nu(1-\beta_i)(\zeta_i - 1)P_{t-1} + \gamma_i \beta_i P_{i,t-1} \Big]. \tag{6}$$

2.1 Model property: the case of independence

The aim of this section is to describe the memory property of the financial time series P_t, in the case of absence of relations between the strategy β_i, adopted by the agent i, and the weight $\omega_{i,t}$ of the agent i at time t.

The following result holds.

Theorem 1. *Given $i = 1, \dots, N$, let β_i be a sample drawn from a random variable β such that*

$$\mathbf{E}[\beta^k] \sim O(c)k^{-1-p} + o(k^{-1-p}) \text{ as } k \to +\infty. \tag{7}$$

Moreover, given $i = 1, \dots, N$, let ζ_i be a sample drawn from a random variable ζ. Let us assume that β and ζ are mutually independent.
Furthermore, suppose that $q > 0$ exists such that

$$(\mathbf{E}[\gamma_i])^{k-1} = \bar{\gamma}^{k-1} \sim k^{-q}, \qquad \text{as } k \to +\infty.$$

Then, for $N \to +\infty$ and $q + p \in [-\frac{1}{2}, \frac{1}{2}]$, we have that P_t has long memory with Hurst exponent given by $H = p + q + \frac{1}{2}$.

Proof. Let L be the time-difference operator such that $LP_{i,t} = P_{i,t-1}$.
By definition of $P_{i,t}$, we have

$$(1 - \gamma_i \beta_i L)P_{i,t} = \nu(1 - \beta_i)\bar{P}_{i,t-1} + \nu(1 - \beta_i)(\zeta_i - 1)P_{t-1}, \tag{8}$$

and then

$$P_{i,t} = \frac{\nu(1 - \beta_i)}{1 - \gamma_i \beta_i L}\bar{P}_{i,t-1} + \frac{\nu(1 - \beta_i)(\zeta_i - 1)}{1 - \gamma_i \beta_i L}P_{t-1}. \tag{9}$$

By the definition of P_t and (9), we have

$$P_t = \sum_{i=1}^{N} \omega_{i,t}\Big[\frac{\nu(1 - \beta_i)}{1 - \gamma_i \beta_i L}\bar{P}_{i,t-1} + \frac{\nu(1 - \beta_i)(\zeta_i - 1)}{1 - \gamma_i \beta_i L}P_{t-1}\Big]. \tag{10}$$

Setting the limit as $N \to \infty$ and by the definition of $\bar{P}$, a series expansion gives

$$P_t = \nu \sum_{k=1}^{\infty} \tilde{\omega}_t P_{t-k} \int_{\mathbf{R}}\int_{\mathbf{R}} (\zeta - 1)(1 - \beta)\beta^{k-1}\bar{\gamma}^{k-1} dF(\zeta, \beta). \tag{11}$$

Since, by hypothesis, β and ζ are mutually independent, with distributions F_1 and F_2 respectively, we have

$$P_t = \nu \sum_{k=1}^{\infty} \tilde{\omega}_t P_{t-k}\bar{\gamma}^{k-1} \int_{\mathbf{R}}\int_{\mathbf{R}} (\zeta - 1)(1 - \beta)\beta^{k-1} dF_1(\zeta) dF_2(\beta) =$$

$$= \nu \sum_{k=1}^{\infty} \tilde{\omega}_t P_{t-k}\bar{\gamma}^{k-1} \int_{\mathbf{R}} (\zeta - 1) dF_1(\zeta) \int_0^1 (1 - \beta)\beta^{k-1} dF_2(\beta) =$$

$$= \nu(\bar{\zeta} - 1) \sum_{k=1}^{\infty} \tilde{\omega}_t P_{t-k} \bar{\gamma}^{k-1} (M_{k-1} - M_k),$$

where M_k is the k-th moment of a random variable satisfying the condition (7). Since

$$\underline{\omega} \sum_{k=1}^{\infty} P_{t-k}(M_{k-1} - M_k) < \sum_{k=1}^{\infty} \tilde{\omega}_t P_{t-k}(M_{k-1} - M_k) < \overline{\omega} \sum_{k=1}^{\infty} P_{t-k}(M_{k-1} - M_k)$$

and

$$M_{k-1} - M_k \sim k^{-p-1}, \tag{12}$$

then, by the hypothesis on the γ_i's, we desume

$$\bar{\gamma}^{k-1}(M_{k-1} - M_k) \sim k^{-q-p-1}. \tag{13}$$

Therefore we have a long memory model $I(d)$ with $d = p + q + 1$ and thus Hurst exponent $H = p + q + \frac{1}{2}$ ([DG1996a], [DG1996b], [G], [GY9] and [E]).

Remark 1. We can use the Beta distribution $B(p, q)$ for defining the random variable β. In fact, if X is a random variable such that $X \sim B(p, q)$, with $p, q > 0$, then X satisfies the relation stated in (7).

Remark 2. In the particular case $\gamma_i = 1$, for each $i = 1, \ldots, N$, the long term memory is allowed uniquely for persistence processes. In this case it results $q = 0$ and, since $p > 0$ by definition, Theorem 1 assures that $H \in (\frac{1}{2}, 1]$.

Remark 3. Structural changes drive a change of the Hurst's parameter of the time series, and thus the degree of memory of the process. In fact, if the chartist calibrating parameter γ_i or the proportionality factor between chartist and fundamentalist, β_i, vary structurally, then the distribution parameters p and q of the related random variables change as well. Therefore, H varies, since it depends on q and p. Furthermore, a drastic change can destroy the stationarity property of the time series. In fact, in order to obtain such stationarity property for P_t, we need that $p+q \in [-1/2, 1/2]$, and modifications of q and/or p must not exceed the range.

Remark 4. The parameters q and p could be calibrated in order to obtain a persistent, anti-persistent or uncorrelated time series.

2.2 Model property: introducing the dependence structure

This section aims to describe the long-run equilibrium properties of financial time series, in the case in which the weights of the investors can drive the forecasts' strategies. The approach we propose allows consideration of the presence of imitative behaviors among the agents. The phenomena of the herding investors is a regularity of financial markets. Since the empirical evidence of crises of the markets, the interests of a wide part of economists have been focused on the analysis of the financial systems fragility. A part of the literature emphasized the relationship between financial crises and weak fundamentals of the economy ([AG], [BER] and [CPR]). A possible explanation of the reasons for the fact that asset prices do not reflect the fundamentals, can be found in the spreading of information among investors, and in the consequent decision to follow a common behavior.

We model the dependence structure allowing the size of the order to change the proportion between fundamentalist and chartist forecasts.

Then, for each weight $\omega_{i,t}$, we consider a function

$$f_{\omega_{i,t}} : D[0,1] \to D[0,1] \text{ such that } f_{\omega_{i,t}}(\beta) = \tilde{\beta}, \qquad \forall i,t. \tag{14}$$

Analogously to the previous section, we formalize a result on the long-run equilibrium properties of the time series P_t in this setting.

Theorem 2. *Given $i = 1, \dots, N$, let β_i be a sampling drawn from a random variable $\beta \in D[0,1]$.*
Fixed $\omega_{i,t}$, let $f_{\omega_{i,t}}$ be a random variable transformation defined as in (14) such that

$$\mathbf{E}[\{f_{\omega_{i,t}}(\beta)\}^k] = \mathbf{E}[\tilde{\beta}^k] \sim O(c)k^{-1-\tilde{p}} + o(k^{-1-\tilde{p}}) \text{ as } k \to +\infty. \tag{15}$$

Moreover, given $i = 1, \dots, N$, let ζ_i be a sample drawn from a random variable ζ, where $\tilde{\beta}$ and ζ are mutually independent.
Furthermore, suppose that $q > 0$ exists such that

$$(\mathbf{E}[\gamma_i])^{k-1} = \bar{\gamma}^{k-1} \sim k^{-q}, \qquad \text{as } k \to +\infty.$$

Then, for $N \to +\infty$ and $q + \tilde{p} \in [-\frac{1}{2}, \frac{1}{2}]$, we have that P_t has long memory with Hurst exponent given by $H = \tilde{p} + q + \frac{1}{2}$.

Proof. The proof is similar to the one given for Theorem 1.

Remark 5. Remark 1 guarantees that the $f_{\omega_{i,t}}$ can transform $X \sim B(p,q)$ in $f_{\omega_{i,t}}(X) \sim B(\tilde{p}, \tilde{q})$. Therefore, the changing of the strategy used by the investors, driven by the weights ω's, can be attained by calibrating the parameters of a Beta distribution.

We use the $B(p,q)$ distribution because of its statistical properties and of the several different shapes that it can assume depending on its parameter values. In the particular case $p = 1$, $q = 1$ is the uniform distribution. If β_i is sampled in accordance with a uniform distribution, then there is no prevailing preference on the strategy, and also between either chartist or fundamentalist approach. If β_i is sampled in accordance with a random variable β, $\beta \sim B(p,p)$, $p > 1$, then this means that the agents opinion agree on mixture parameter values close to the mean of β. If the distribution is U-shaped, this means that there are two more agreeable strategies.

3 Conclusions

This paper has shown how to consider the weight of a market trader in a micro-economic model that allows the theoretical statement of the degree of long memory in data. Since $H = 1/2$ is taken into account in the theoretical model, the long-run equilibrium properties of uncorrelated processes represents a particular case. Therefore, the model encompasses a wide range of processes. This approach allows the avoidance of time-expensive numerical calibration and allows the use of the model also for explaining the weight distribution on high-frequency trading. A further analyses has been carried out on a more detailed correspondence between the group size and the trader strategy, in both cases of dependence and independence.

A bayesian statistic approach, to develop the analysis of the dependence structure between size and strategies, can be used. We leave this topic to future research.

References

[AG] Allen, F., Gale, D.: Optimal Financial Crises. Journal of Finance, **53**, 1245–1284 (1998)

[AIV] Ausloos, M., Ivanova, K., Vandervalle, N.: Crashes: symptoms, diagnoses and remedies. In: Takayasu, H. (ed) Empirical sciences of financial fluctuations. The advent of econophysics. Springer Verlag, Berlin (2002)

[BPS] Bayraktar, E., Poor, H.V., Sircar, K.R.: Estimating the fractal dimension of the S&P 500 index using wavelet analysis. International Journal of Theoretical and Applied Finance, **7**, 615–643 (2004)

[Be] Bershadskii, A.: Multifractal diffusion in NASDAQ. J. Phys. A: Math. Gen. **34**, L127–L130 (2001)

[BP] Bianchi, S., Pianese, A.: Scaling Laws in Stock Markets. An Analysis of Prices and Volumes. Preprint

[BER] Burnside, C., Eichenbaum, M., Rebelo, S.: On the Fundamentals of Self-Fulfilling Speculative Attacks. University of Rochester, Center for Economic research WP468 (2001)

[CR] Cerqueti, R., Rotundo, G.: Microeconomic modeling of financial time series with long term memory. Proceedings of IEEE International Conference on Computational Intelligence for Financial Engineering (CIFEr2003), Hong Kong, 191–198 (2003)

[CPR] Corsetti, G., Pesenti, P., Roubini, N.: What Caused the Asian Currency and Financial Crisis? Part I: A Macroeconomic Overview. Mimeo (1999)

[DEG] Ding, Z., Engle, R.F., Granger, C.W.J.: A long memory property of stock market returns and a new model. Journal of Empirical Finance, **1**, 83–106, (1993)

[DG1996a] Ding, Z., Granger, C.W.J.: Varieties of long memory models. Journal of Econometrics, **73**, 61–77 (1996)

[DG1996b] Ding, Z., Granger, C.W.J.: Modelling volatility persistence of speculative returns: a new approach. Journal of Econometrics, **73**, 185–215 (1996)

[F] Foellmer, H.: Random economies with many interacting agents. Journal of Mathematical Economics, **1**, 51–62 (1974)

[FHK] Foellmer, H., Horst, U., Kirman, A.: Equilibria in financial markets with heterogeneous agents: a probabilistic perspective. Journal of Mathematical Economics, **41**, 123–155 (2005)

[G] Granger, C.W.J.: Long memory relationships and the aggregation of dynamic models. Journal of Econometrics, **14**, 227–238 (1980)

[GY] Granger, C.W.J., Yoyeux, R.: An Introduction to Long Memory Time series Models and Fractional differencing. J. Time Series Anal., **1**, 15–29 (1980)

[K1993] Kirman, A.: Ants, rationality and recruitment, Quaterly Journal of Economics, **108**, 137–156 (1993)

[K1991] Kirman, A.: Epidemics of opinion and speculative bubbles in financial markets. In: Taylor, M. (ed) Money and Financial Markets. Macmillan, London (1991)

[KT] Kirman, A.P., Teyssiére G.: Microeconomic Models for Long-Memory in the Volatility of Financial Time Series, Preprint GREQAM DT 00A3 (2000)

[L] Lo, A.W.: Long memory in stock market prices. Econometrica, **59**, 1279–1313 (1991)

[MS] R. Mantegna, R., Stanley, H.E.: An Introdution to Econophysics, Cambridge Univ. Press (2000)

[RB] Reka, A., Barabasi, A.L.: Statistical mechanics of complex networks. Reviews of Modern Physics, **74**, 47–97 (2002)

[R] Rotundo, G.: Microeconomic model switching in bubble dynamics, Preprint

[SSL] Schmitt, F., Schertzer, D., Lovejoy, S.: Multifractal fluctuations in finance, Int. J. Appl. Fin., **3**, 361–364 (2000)

[VDA] Vandervalle, N., D'hulst, R., Ausloos, M.: Phase segregation in binary sandpiles on fractal bases. Phys. Rev. E, **59**, 631–635 (1999)

A Liability Adequacy Test for Mathematical Provision*

Rosa Cocozza, Emilia Di Lorenzo, Abina Orlando and Marilena Sibillo

Summary. This paper deals with the application of the Value at Risk of the mathematical provision within a fair valuation context. Through the VaR calculation, the estimate of an appropriate contingency reserve is connected to the predicted worst case additional cost, at a specific confidence level, projected over a fixed accounting period. The numerical complexity is approached by means of a simulation methodology, particularly suitable also in the case of a large number of risk factors.

Key words: Value at Risk; Life insurance; Quantile Reserve; Solvency, Fair Value.

1 Liability Adequacy Test and Contingency Reserve

The International Financial Reporting Standard 4 (IFRS4) for insurance contracts requires a liability adequacy test, whose essentials are given in IFRS4.15, which states: *an insurer shall assess at each reporting date whether its recognized insurance liabilities are adequate, using current estimates of future cash flow under its insurance contracts. If that assessment shows that the carrying amount of its insurance liabilities (less related deferred acquisition costs and related intangible assets [...]) is inadequate in the light of the estimated future cash flows, the entire deficiency shall be recognized in profit or loss.*

As a consequence, an insurer has to check the adequacy of the "recognized amount" with the "net carrying amount". According to the Basis for Conclusion (BC100), *the comparison is made at the level of a portfolio of contracts that are subject to broadly similar risks and managed together as a portfolio.* Since the IFRS4 is a stepping stone to a wider revision of insurance reporting topics, it is beyond its scope to create a detailed accounting regime for an insurance contract; therefore, among other issues relating to liability adequacy, it does not specify whether or how the cash flows are discounted to reflect the time value of money or adjusted for risk and uncertainty.

* Although the paper is the result of a common study, Section 1 is credited to R. Cocozza and Sections 2 and 3 are credited to E. Di Lorenzo, A. Orlando and M. Sibillo

Cira Perna, Marilena Sibillo (Eds.): Mathematical and Statistical Methods in Insurance and Finance

The International Accounting Standard 37 (IAS37) states that *the amount recognized as a provision shall be the best estimate of the expenditure required to settle the present obligation at the balance sheet date*, where the specific best estimate is the amount that an entity would rationally pay to settle the obligation at the balance sheet date, or to transfer it to a third party at that time (IAS37.36), where best estimates are determined by judgment of the management of the entity, supplemented by experience of similar transactions and, in some cases, reports from independent experts (IAS 37.35).

Combining the IFRS4 prescriptions with the fair value specifications and hierarchy, the fair value of an insurance liability results to be not only an "estimated value" but a present value. Consistently, its fair value can be defined as the net present value of the residual debt towards the policyholders evaluated at current interest rates and, eventually, at current mortality rates. In one sense, this is marking to market; though it is marking to model. This implies discounting the cash flow using the government yield curve plus an appropriate spread. The spread should reflect an appropriate market premium for risks other than interest rate risk, which could be the option adjusted spread of the replicating portfolio if existent or a proxy for it. If there is an amendment in the evaluation criteria from one reporting date to another (according to current market yields and/or mortality rates) there is a possible change in the value of the reserve according to the application of a more stringent or flexible criterion. This may turn into a proper fair valuation risk (see [CDL]). The introduction of an accounting policy involving re-measuring designated insurance liabilities consistently in each period to reflect current market interest rates (and, if the insurer so elects, other current estimates and assumptions) implies that the fair value of the mathematical provision is an "adjustable present value". The variation of such evaluated price gives rise to a fair valuation risk and opens the path to a quantification of contingency provisions.

As far as the valuation is concerned, the definition of the net carrying amount is a "current net present value", whose result is subject to a yield curve and, if eligible, to a mortality table. Setting apart the question of mortality rates, the yield curve normally changes during time and, consistently, modifies the corresponding present value, giving rise to an assortment of future prices. Therefore, the net carrying amount can be defined as "the expected value of a distribution of provision values, set by the yield curve shifts", whose general calculation is given by Equation (1) (cf. Section 2). This is what IAS37 requires. Yet, if there is a distribution of current values (spot prices), there is also a distribution of future values (forward prices). This implies that the "selection" of a current value gives rise to an expected future value, which is in turn connected to a distribution of values. If it is possible to model the distribution of future values (cf. Section 2), at the end of each reporting period it is possible to define not only the net current present value and its variance, but also the corresponding expected future value and the variance of the future value distribution. This last information can be used to define the future expected value and, through value-at-risk calculation, the predicted worst case additional cost at a specific confidence level over the next accounting period (cf. Section 3). This last value can be treated as a contingency reserve account, even in the form of a solvency reserve.

2 A Solvency Perspective via the Quantile Reserve

In a fair valuation context, let us introduce two probability spaces $(\Omega, \mathcal{F}', P')$ and $(\Omega, \mathcal{F}'', P'')$, where $\mathcal{F}'$ and $\mathcal{F}''$ are the σ-algebras containing, respectively, the *financial events* and the *life duration events*. Given the independence of the mortality fluctuations on interest rate randomness, we denote by $(\Omega, \mathcal{F}, P)$ the probability space canonically generated by the preceding two. $\mathcal{F}$ contains information flow about both mortality and financial history, represented by the filtration $\{\mathcal{F}_k\} \subseteq \mathcal{F}(\mathcal{F}_k = \mathcal{F}'_k \cup \mathcal{F}''_k$ with $\{\mathcal{F}'_k\} \subseteq \mathcal{F}'$ and $\{\mathcal{F}''_k\} \subseteq \mathcal{F}''$).

Assuming a frictionless market, with continuous trading, no restrictions on borrowing or short-sales, the zero-bond and the stocks being both infinitely divisible, the fair value of the reserve at time t of a portfolio of life insurance contracts (with obvious meaning of the symbol $\mathcal{F}_t$) is given by

$$V_t = E\left[\sum_{r>t} CF_r v(t,r)/F_t\right], \tag{1}$$

where $v(t,r)$ is the present value at time t of one monetary unit due at time r, CF_r is the net cash flow at time r, and E represents the expectation under the risk-neutral probability measure, whose existence derives from well known results, based on the completeness of the market.

With respect to the demographic risk context, given that the demographic valuation is not supported by the hypotheses of completeness of the market, the current valuation can be represented by means of the expectation consistently with the best prediction of the demographic scenario. In a general perspective, a fair valuation procedure involves the latest information on the two main factors bringing risk to the business, interest rates and mortality (see [CDLOS]).

Equation (1) can be easily specialized in the case of a portfolio of different life annuities with benefits payable at the beginning of each period. In this case we split the portfolio into m homogeneous sub-portfolios characterized by common aspects, that is age at issue, policy duration, payments, deferment period, etc.

Let us introduce the following notations:

$$n = \text{maximum term for all contracts,}$$

$$S_{i,r} = \text{number of survivors at time } r \text{ of the insureds of the } i\text{-th group,}$$

$$L_i = \text{constant annual benefit for each contract of the } i\text{-th group,}$$

$$P_{i,r} = \text{premium paid at time } r \text{ for each contract of the } i\text{-th group,}$$

$$T_i = \text{deferment period } (T_i = 0, 1, \ldots),$$

$$\tau_i = \text{premium payment period } (0 \leq \tau_i < T_i).$$

Hence (1) becomes

$$V_t = E\left[\sum_{r=t+1}^{m} \sum_{i=1}^{m} \left[S_{i,r}\left(L_i \mathbf{1}_{(n_i \geq r)\wedge(r \geq T_i)} - P_{i,r} \mathbf{1}_{(n_i \geq r)\wedge(r < \tau_i)}\right)\right] v(t,r)/F_t\right], \tag{2}$$

where the indicator function $\mathbf{1}_{(n_i \geq r) \wedge (r \geq T_i)}$ takes the value 1 if $n_i \geq r$ and $r \geq T_i$, otherwise 0, whilst the indicator function $\mathbf{1}_{(n_i \geq r) \wedge (r < \tau_i)}$ takes the value 1 if $n_i \geq r$ and $r < \tau_i$, otherwise 0.

Equation (2) is framed in a *forward* perspective, within a current valuation provided at the initial position 0. Analogously this formula can be re-interpreted in a *spot* perspective, according to a year by year valuation, that is providing the current value of the reserve at the end of the year, valued at the beginning of the year itself (cf. [CDLOS]).

In a solvency assessment perspective, models involving the so-called *quantile reserve* play a fundamental role because of their specific links to the *Value-at-Risk*.

Indicating by $R(t)$ the financial position at time t, that is, in this case, the stochastic mathematical provision of a portfolio of contracts, the quantile reserve at confidence level α $(0 < \alpha < 1)$, is the value $R^*_\alpha(t)$ implicitly defined by the following equation:

$$P\{R(t) > R^*_\alpha(t)\} = 1 - \alpha. \tag{3}$$

Moreover, considering the time interval $[t, t+h]$ and the financial positions at its extremes, say $r(t)$ and $R(t+h)$ respectively, the potential periodic loss is defined as (Teugels et al. 2002):

$$L = r(t) - R(t+h), \tag{4}$$

therefore at confidence level α, the *Value-at-Risk* $VaR(\alpha)$ is given by:

$$P\{L > VaR(\alpha)\} = 1 - \alpha \Leftrightarrow VaR(\alpha) = F^{-1}(\alpha), \tag{5}$$

F being the cumulative distribution function of $R(t)$.

3 A Simulative Application

In this section we propose a simulation procedure to quantify the VaR of two homogeneous portfolios of deferred life annuities. For the sake of simplicity, the valuation of the financial instruments composing the ZCB portfolio will be made assuming a term structure of interest rates based on a square root CIR process:

$$dr_i = -\alpha(r_i - \mu)dt + \sigma\sqrt{r_i}dW_t,$$

with α and σ positive constants, μ the long term mean and W_t a Wiener process.

Referring to the mortality scenario, in a fair valuation estimating framework, we consider a marking-to-model system for the demographic quantities. We assume that the best prediction for the time evolution of the surviving phenomenon is represented by a fixed set of survival probabilities, estimated taking into account the improving trend of mortality rates (best estimate).

For each portfolio, we simulate 100,000 values of the potential periodic loss L defined in Section 2. The simulated $\{L(j)\}$), $j = 1, 2, \ldots, 100{,}000$ can be treated as a sample from a normal distribution (cf. [CDLOS]), which we use to estimate the VaR. In Fig.1 and Table 1 we present the results obtained for a portfolio of

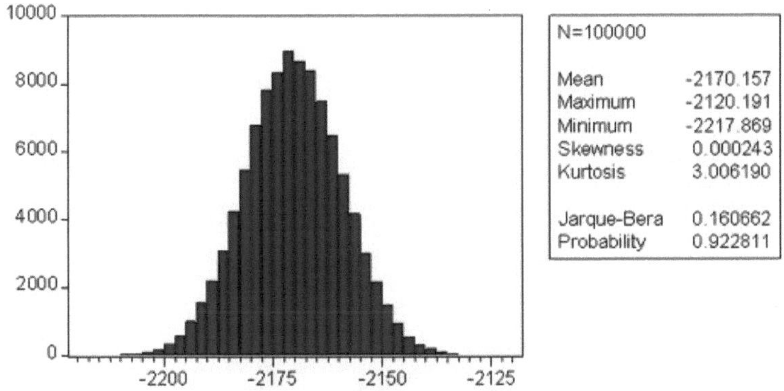

Fig. 1. Histogram of L function at time $T = 2$. 1000 deferred life annuities. Age at issue 30. CIR process $\alpha = 0.026$, $\mu = 0.045$, $\sigma = 0.00052$; initial value $r_0 = 0.0172$ (Cocozza et al., 2004). Mortality: Italian Male RG48.

1000 deferred 15-years temporary unitary life annuities (deferment period $T = 5$), for policyholders aged 30 at issue. Periodic premiums are paid at the beginning of each year of the deferment period. In particular, Fig. 1 shows the histogram and the characteristic values of the Loss function simulated distribution.

Referring to (2) and (4) of Section 2, we compute the VaR at time $t = 2$, during the deferment period. We frame the calculation within a spot perspective, according to a year by year valuation.

Under the above hypothesis about survival and rates, in Fig. 2 and Table 2 we present the results obtained for a portfolio of 1000 deferred 10-years temporary unitary life annuities (deferment period $T = 3$), for policyholders aged 40 at issue, considering the VaR at time $t = 7$.

Recalling (5), the two VaR tables supply the maximum future net carrying amount of the final reserve within a set confidence interval. In the first case (Table 1), the value is negative because it sets an expense since the reserve is in a growing phase, that is to say $R(t+h) > r(t)$. In the second example (Table 2), the value is positive because the reserve is in a decreasing phase, i.e. $R(t+h) < r(t)$. These values are a probabilistic projection of future difference between initial and final reserve and can be useful for the quantification of the contingency reserve.

Table 1. Value at risk $T = 2$. 1000 deferred life annuities. Age at issue 30.

Confidence level	*VaR*
99%	−2144.198
95%	−2151.7156

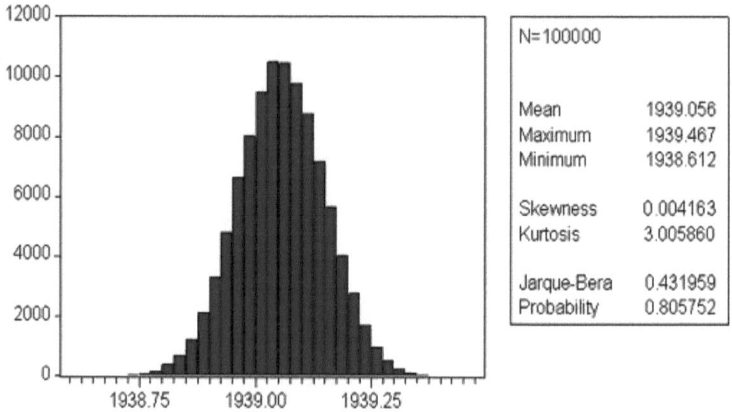

Fig. 2. Histogram of L function at time $T = 7$. 1000 deferred life annuities. Age at issue 40. CIR process $\alpha = 0.026$, $\mu = 0.045$, $\sigma = 0.00052$; the initial value $r_0 = 0.0172$ (Cocozza et al., 2004). Mortality: Italian Male RG48.

Table 2. Value at risk $T = 7$. 1000 deferred life annuities. Age at issue 40.

Confidence level	*VaR*
99%	1939.277
95%	1939.213

References

[CKLS] Chan, K.C., Karolyi, A.G., Longstaff, F.A., Sanders, A.B.: An Empirical Comparison of Alternative Models of the Short-Term Interest Rate. The Journal of Finance, **47**, 1209–1227 (1992)

[CM] Cleur, E.M., Manfredi, P.: One dimensional SDE models, low order numerical methods and simulation based estimation: a comparison of alternative estimators. Computational Economics, **13(2)**, 177–197 (1999)

[CDL] Cocozza, R., Di Lorenzo, E.: Solvency of life insurance companies: methodological issues. Journal of Actuarial Practice, **13**, 81–101 (2005)

[CDLOS] Cocozza, R., Di Lorenzo, E., Orlando, A., Sibillo, M.: The VaR of the mathematical provision: critical issue. In: Proceedings of the International Conference of Actuaries http://papers.ica2006.com/Papiers/2002/2002.pdf (2006)

[CDLS1] Cocozza, R., Di Lorenzo, E., Sibillo, M.: Risk Profiles of Life Insurance Business: quantitative analysis in a managerial perspective. In: Atti del convegno Metodi Matematici e Statistici per l'Analisi dei Dati Assicurativi e Finanziari - MAF2004, 81–86 (2004)

[CDLS2] Cocozza, R., Di Lorenzo, E., Sibillo, M.: Methodological problems in solvency assessment of an insurance company. Investment Management and Financial Innovations, **2**, 95–102 (2004)

[CDLS3] Cocozza, R., Di Lorenzo, E., Sibillo, M.: Life insurance risk indicators: a balance sheet approach. In: Proceedings of the VIII International Congress on Insurance: Mathematics & Economics. http://www.ime2004rome.com/fullpapers/Cocozza%20PAPE.pdf (2004)

[CoDLS1] Coppola, M., Di Lorenzo, E., Sibillo, M.: Risk Sources in a Life Annuity Portfolio: Decomposition and Measurement Tools. Journal of Actuarial Practice, **8**, 43–61 (2000)

[CoDLS2] Coppola, M., Di Lorenzo, E., Sibillo, M.: Further Remarks on Risk Sources Measuring in the Case of a Life Annuity Portfolio. Journal of Actuarial Practice, **10**, 229–242 (2002)

[R] Rubinstein, Y.R.: Simulation and Monte Carlo method. Wiley & Sons, Chichester (1981)

[TZ] Talay, D., Zheng, Z.: Quantile of the Euler scheme for diffusion processes and financial application. Mathematical Finance, **13**, 187–199 (2003)

[TS] Teugels, J.L., Sundt, B. Editors: Encyclopedia of Actuarial Science. Wiley & Sons, Chichester (2004)

Iterated Function Systems, Iterated Multifunction Systems, and Applications*

Cinzia Colapinto and Davide La Torre

Summary. In the first part of the paper we recall the theory of iterated function systems and iterated multifunction systems. In the second part we show some applications in economics, statistics and finance.

Key words: Iterated function systems; Iterated multifunction systems; Fractal approximations; Estimation; Stochastic processes.

1 Introduction

Iterated Function Systems (IFS) are a very nice way to formalize the notion of self-similarity or scale invariance of some mathematical objects. Hutchinson ([Hu81]) and Barnsley and Demko ([Ba85, Ba89 and Ba97]) showed how systems of contractive maps with associated probabilities can be used to construct fractal sets and measures. Many models in decision theory, economics, finance, image processing and statistics, involve these types of dynamic systems (see [Ia05, Ia051, Ia06, La06, La061 and Vr99]). The extensions of IFS-type methods is an ongoing research program on the construction of appropriate IFS-type operators, or *generalized fractal transforms* (GFT), over various spaces, i.e., function spaces and distributions, vector-valued measures, integral transforms, wavelet transforms, set valued functions and set valued measures (see [Fo98, Fo981, Fo99, Hu81, Ku07, La061 and Me97). The action of a generalized fractal transform $T : X \to X$ on an element u of the complete metric space (X, d) can be summarized in the following steps. It produces a set of N spatially-contracted copies of u and then modifies the values of these copies by means of a suitable range-mapping. Finally, it recombines them using an appropriate operator in order to get the element $v \in X$, $v = Tu$. In all these cases, under appropriate conditions, the fractal transform T is a contraction and thus Banach's fixed point theorem guarantees the existence of a unique fixed point $\bar{u} = T\bar{u}$. The inverse

* This work was written during a research visit to the Department of Applied Mathematics of the University of Waterloo, Canada.

Cira Perna, Marilena Sibillo (Eds.): Mathematical and Statistical Methods in Insurance and Finance

problem is a key factor for applications: given $T : X \to X$ a point-to-point contraction mapping and a "target" element $y \in X$, we look for a contraction mapping T with fixed point $\bar{x}$ such that $d(y, \bar{x})$ is as small as possible. In practical applications however, it is difficult to construct solutions for this problem and one relies on the following simple consequence of Banach's fixed point theorem, known as "collage theorem", which states that $d(y, \bar{x}) \le \frac{1}{1-c} d(y, Ty)$ (c is the contractivity factor of T). Instead of trying to minimize the error $d(y, \bar{x})$, one looks for a contraction mapping T that minimizes the *collage error* $d(y, Ty)$.

Iterated Multifunction Systems (IMS) are a generalization of IFS, from a standard point-to-point contraction mapping to a set-valued operator. IMS operators T are defined by the parallel action of a set of contractive multifunctions T_i. Under suitable conditions T is contractive, implying the existence of a fixed-point $\bar{x}$ which is a solution of the fixed point inclusion $\bar{x} \in T\bar{x}$. The multifunction T satisfies the following contractivity condition: there exists a $c \in [0, 1)$ such that $d_h(Tx, Ty) \le cd(x, y)$ for all $x, y \in X$, where d_h denotes the Hausdorff metric between sets Tx and Ty. From a Covier and Naylor theorem, if T is contractive in the above sense, then a fixed point $\bar{x} \in X$ exists such that $\bar{x} \in T\bar{x}$. Note that $\bar{x}$ is not necessarily unique. The set of fixed points of T, to be denoted as X_T, play an important role in application. A corollary of the Covier-Naylor theorem, based on projections onto sets, is a method to construct solutions to the fixed point equation $x \in Tx$, essentially by means of an iterative method that converges to a point $x \in X_T$. For IMS it is possible to show two results that can be viewed as multifunction analogues of those that apply when T is a contractive point-to-point mapping (in which case Banach's fixed point theorem applies), namely a continuity property of fixed point sets X_T and a collage theorem for multifunctions (see [Ku07]).

2 Iterated Function Systems (IFS)

Let $d(x, y)$ be the Euclidean distance and $\mathcal{H}(X)$ denote the space of all compact subsets of X and $d_h(A, B)$ the Hausdorff distance between A and B, that is $d_h(A, B) = \max\{\max_{x \in A} d'(x, B), \max_{x \in B} d'(x, A)\}$, where $d'(x, A)$ is the usual distance between the point x and the set A, i.e. $d'(x, A) = \min_{y \in A} d(x, y)$. We will write $h(A, B) = \max_{x \in A} d'(x, B)$. It is well known that the space $(\mathcal{H}(X), d_h)$ is a complete metric space if X is complete. First of all we introduce the idea of an iterated function system. (X, d) denotes a complete metric space, typically $[0, 1]^n$. Let $\mathbf{w} = \{w_1, \cdots, w_N\}$ be a set of contraction maps $w_i : X \to X$, to be referred to as an N-map IFS. Let $c_i \in [0, 1)$ denote the contraction factors of the w_i and define $c = \max_{1 \le i \le N} c_i \in [0, 1)$. As before, we let $\mathcal{H}(X)$ denote the set of non-empty compact subsets of X and h the Hausdorff metric. Associated with the IFS maps, w_i is a set-valued mapping $w : \mathcal{H}(X) \to \mathcal{H}(X)$ the action of which is defined to be $w(S) = \bigcup_{i=1}^{N} w_i(S)$, $S \in \mathcal{H}(X)$, where $w_i(S) := \{w_i(x), x \in S\}$ is the image of S under w_i, $i = 1, 2, \cdots, N$. It is a standard result that w is a contraction mapping on $(\mathcal{H}(X), d_h)$, that is $d_h(w(A), w(B)) \le cd_h(A, B)$, $A, B \in \mathcal{H}(X)$. Consequently, a unique set $A \in \mathcal{H}(X)$ exists, such that $w(A) = A$, the so-called *attractor* of the

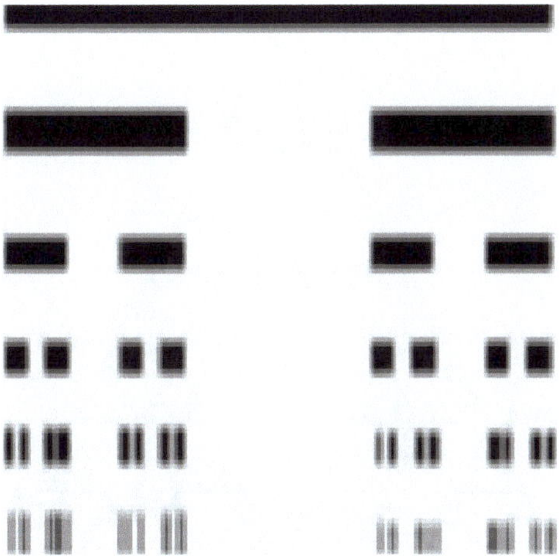

Fig. 1. Cantor set.

IFS. The equation $A = w(A)$ obviously implies that A is self-tiling, i.e. A is a union of (distorted) copies of itself. Moreover, for any $S_0 \in \mathcal{H}(X)$, the sequence of sets $S_n \in \mathcal{H}(X)$ defined by $S_{n+1} = w(S_n)$ converges in Hausdorff metric to A.

Example 1. $X = [0, 1]$, $N = 2$: $w_1(x) = \frac{1}{3}x$, $w_2(x) = \frac{1}{3}x + \frac{2}{3}$. Then the attractor A is the ternary Cantor set C on $[0, 1]$ (see Fig. 1).

Example 2. Fig. 2 represents the Sierpinski gasket. This set is the attractor of the IFS $\{w_1, w_2, w_3\}$ acting on $\mathbb{R}^2$, where $w_i(x) = \frac{1}{2}(x - P_i) + P_i$, $i = 1, 2, 3$ and the points P_1, P_2 and P_3 are the vertices of the outer triangle.

Let $\mathcal{M}(X)$ be the set of probability measures on $\mathcal{B}(X)$, the σ-algebra of Borel subsets of X where (X, d) is a compact metric space. In the IFSs literature, the following *Monge-Kantorovich* metric plays a crucial role

$$d_M(\mu, \nu) = \sup_{f \in \mathrm{Lip}(X)} \left\{ \int_X f d\mu - \int_X f d\nu \right\}, \quad \mu, \nu \in \mathcal{M}(X), \tag{1}$$

where $\mathrm{Lip}(X) = \{f : X \to \mathbb{R}, |f(x) - f(y)| \leq d(x, y), x, y \in X\}$ thus $(\mathcal{M}(X), d_M)$ is a complete metric space (see [Hu81]). We denote by $(\mathbf{w}, \mathbf{p})$ an *N-maps contractive IFS on X with probabilities* or simply an *N-maps IFS*, that is, a set of N affine contraction maps $\mathbf{w}$ with associated probabilities $\mathbf{p} = (p_1, p_2, \ldots, p_N)$, $p_i \geq 0$, and $\sum_{i=1}^N p_i = 1$. The IFS has a contractivity factor defined as $c = \max_{1 \leq i \leq N} |b_i| < 1$. Consider the following (usually called *Markov*) operator $M : \mathcal{M}(X) \to \mathcal{M}(X)$ defined as $M\mu = \sum_{i=1}^N p_i \mu \circ w_i^{-1}$, $\mu \in \mathcal{M}(X)$, where w_i^{-1} is the inverse function of w_i and $\circ$ stands for the composition. In [Hu81] it was shown that M is a contraction mapping on $(\mathcal{M}(X), d_M)$ i.e. for all $\mu, \nu \in \mathcal{M}(X)$, $d_M(M\mu, M\nu) \leq c d_M(\mu, \nu)$. Thus, a unique measure $\bar{\mu} \in \mathcal{M}(X)$ exists, the *invariant measure* of the IFS, such that $M\bar{\mu} = \bar{\mu}$ by Banach theorem.

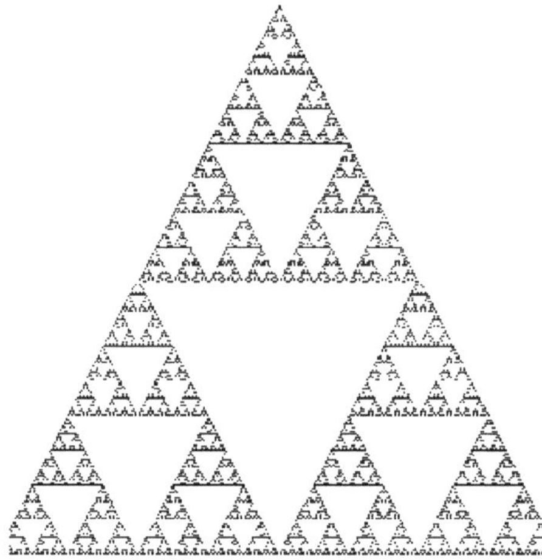

Fig. 2. Sierpinski gasket.

3 Iterated Multifunction Systems

As an extension of IFS, consider a set of $T_i : X \rightrightarrows X$ of multifunctions where $i \in 1 \ldots n$ and $T_i x \in \mathcal{H}(X)$ for all i. We now construct the multifunction $T : X \rightrightarrows X$ where $Tx = \bigcup_i T_i x$. Suppose that the multifunctions T_i are contractions with contractivity factor $c_i \in [0, 1)$, that is, $d_h(T_i x, T_i y) \leq c_i d(x, y)$. From the Covier-Nadler theorem cited earlier, a point $\bar{x} \in T\bar{x}$ exists. Now, given a compact set $A \in \mathcal{H}$, consider the image $T(A) = \bigcup_{a \in A} Ta \in \mathcal{H}(X)$. Since $T : (X, d) \to (\mathcal{H}(X), d_h)$ is a continuous function, then $T(A)$ is a compact subset of $\mathcal{H}(X)$. So we can build a multifunction $T^* : \mathcal{H}(X) \rightrightarrows \mathcal{H}(X)$ defined by $T^*(A) = T(A)$ and consider the Hausdorff distance on $\mathcal{H}(X)$, that is given two subsets $A, B \subset \mathcal{H}(X)$ we can calculate

$$d_{hh}(A, B) = \max\{\sup_{x \in A} \inf_{y \in B} d_h(x, y), \sup_{x \in A} \inf_{y \in B} d_h(x, y)\}. \tag{2}$$

We have that $T^* : \mathcal{H}(X) \rightrightarrows \mathcal{H}(X)$ and $d_{hh}(T^*(A), T^*(B)) \leq c d_h(A, B)$. Now, given a point $x \in X$ and a compact set $A \subset X$, we know that the function $d(x, a)$ has at least one minimum point $\bar{a}$ when $a \in A$. We call $\bar{a}$ the *projection* of the point x on the set A and denote it as $\bar{a} = \pi_x A$. Obviously $\bar{a}$ is not unique but we choose one of the minima. We now define the following projection function P associated with a multifunction T defined as $P(x) = \pi_x(Tx)$. We therefore have the following result ([Ku07]).

Theorem 1. *Let (X, d) be a complete metric space and $T_i : X \to \mathcal{H}(X)$ be a finite number of contractions with contractivity factors $c_i \in [0, 1)$. Let $c = \max_i c_i$. Then*

1. *For all compact $A \subset X$ a compact subset $\bar{A} \subset X$ exists such that $A_{n+1} = P(A_n) \to \bar{A}$ when $n \to +\infty$.*
2. *$\bar{A} \subset \bigcup_i T_i(\bar{A})$.*

As in the previous section, let $\mathcal{M}(X)$ be the set of probability measures on $\mathcal{B}(X)$ and consider the complete metric space $(\mathcal{M}(X), d_M)$. Given a set of multifunctions $T_i : X \to X$ with associated probabilities p_i, one can now consider generalized Markov operators on $\mathcal{M}(X)$.

4 Applications

In this section we show some applications of IFS in economics, statistics and finance.

Example 3. (Random dynamical systems in economics) Many questions in economics and finance lead to random discrete dynamical systems $y_{t+1} = w_i(y_t)$ where the map w_i is contractive and it is chosen in a given set with probability p_i. Let us consider, for instance, a macroeconomic system and suppose that the aggregate demand D_t is equal to the income Y_t, for all $t \in \mathbb{N}$. In a closed economy in which the government spending G is equal to 0, the aggregate demand is the sum of consumptions C_t and investments I_t. The quantity of consumption C_t is a function of the income of the previous year by a linear relation as $C_t = \alpha Y_{t-1}$, where the stochastic coefficient α can take values in a given set $C = \{\alpha_1, \alpha_2, \ldots, \alpha_m\} \subset (0, 1)$ with probability p_i, $\sum_{i=1}^{m} p_i = 1$. If the level of investments is constant, that is $I_t = I_0$ (the initial investment level) for all $t \in \mathbb{N}$, the stochastic growth model of the income is described by the system of equations $Y_{t+1} = \alpha Y_t + I_0$ with $P(\alpha = \alpha_i) = p_i$, $i = 1 \ldots m$. In this context, the inverse problem consists of finding the parameters of the model (economical parameters) which allow fixed goals of economic policy to be reached.

Example 4. (Fractal estimation of distribution and density functions) The fractal estimator of a distribution function F is thought as the fixed point of a contractive operator T defined in terms of a vector of parameters p and a family of affine maps $\mathcal{W}$ which can both depend on the sample $(X_1, X_2, \ldots, X_n)$. Given $\mathcal{W}$, the problem consists of finding a vector p such that the fixed point of T is "sufficiently closed" to F. Let $X_1, X_2, \ldots, X_n$ be an i.i.d. sample drawn from a random variable X with unknown distribution function F with compact support $[\alpha, \beta]$. The empirical distribution function (e.d.f.) $\hat{F}_n(x) = \frac{1}{n} \sum_{i=1}^{n} \chi(X_i \leq x)$ is a commonly used estimator of the unknown distribution function F (here χ is the indicator function). Theoretical results prove that a fractal estimator of F is asymptotically equivalent to the empirical distribution function (EDF) estimator. For well behaved distribution functions F and for a particular family of so-called wavelet maps the IFS estimators can be dramatically better than the empirical distribution function in the presence of missing data (see Figs. 3 and 4). The idea of *inverse approach* is to determine p by solving a constrained quadratic optimization problem built in terms of sample moments. The nature of affine maps allow the Fourier transform of F to be easily derived and, when available, an explicit formula for the density of F via anti Fourier transform. In this way, given $\mathcal{W}$ and p we have at the same time estimators for the distribution, characteristic and density functions (for details, see [Ia05] and [Ia051]).

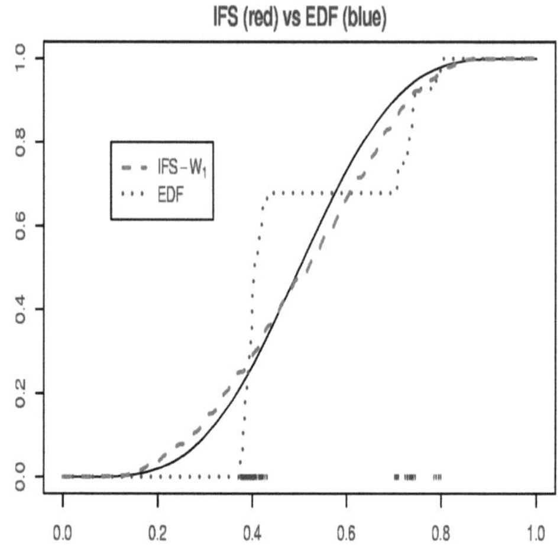

Fig. 3. IFS approximations of distribution functions.

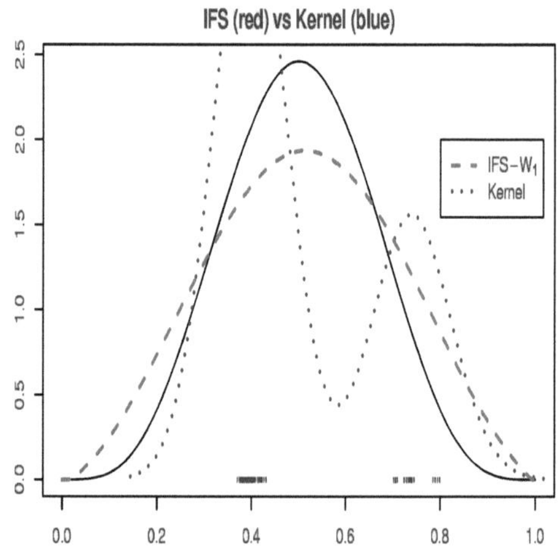

Fig. 4. IFS approximations of density functions.

Example 5. (Fractal approximations of stochastic processes) Let $(\Omega, \mathcal{F}, p)$ be a probability space and $\mathcal{F}_t$ be a sequence of σ algebras such that $\mathcal{F}_t \subset \mathcal{F}$. Let $X(w, t) : \Omega \times [0, 1] \to \mathbb{R}$ be a stochastic process, that is a sequence of random variables $\mathcal{F}_t$ adapted (that is each variable $X(w, t)$ is $\mathcal{F}_t$ measurable). Given $w \in \Omega$ a trajectory of the process is the function $X(w, t) : [0, 1] \to \mathbb{R}$. For a fixed $w \in \Omega$, the trajectory $v(t) = X(t, w)$ is an element of $L^2([0, 1])$. Several methods are currently available to simulate paths of the Brownian motion. In particular, paths of the BM can be simulated using the properties of the increments of the process like in the Euler scheme, or as the limit of a random walk or via L^2 decomposition like

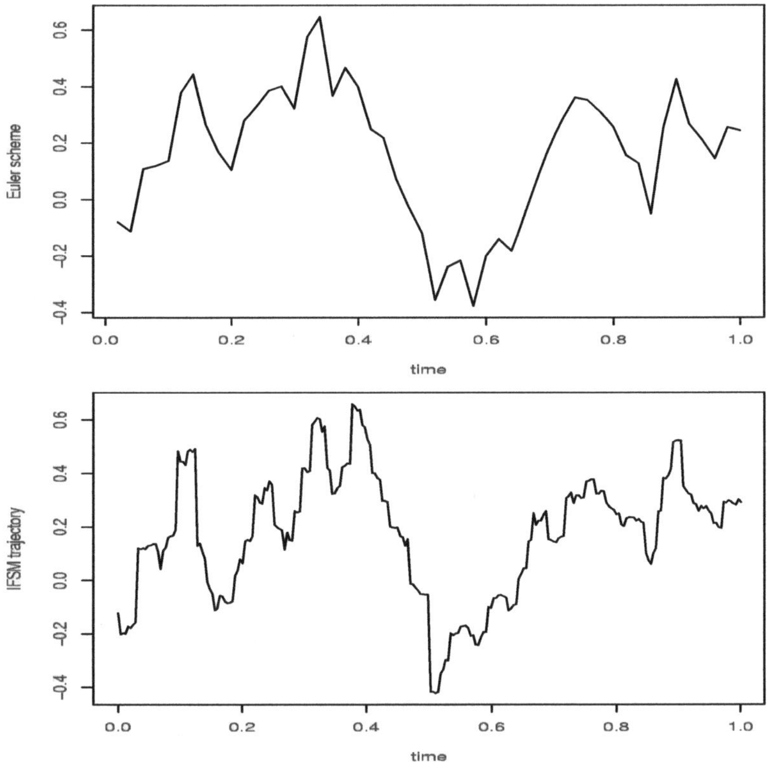

Fig. 5. IFS simulation of Brownian motion.

the Kac- Siegert/Karnounen-Loeve series. IFS operators can be used to simulate trajectories. The resulting simulated trajectories are self-affine, continuous and fractal by construction. This fact produces more realistic trajectories than other schemes in the sense that their geometry is closer to the one of the true BM's trajectories. If $X(w,t)$ is a trajectory of the stochastic process, the inverse approach consists of finding the parameters of the IFSM such that $X(w,t)$ is the solution of the equation $X(w,t) = TX(w,t)$ for e.g. $w \in \Omega$. Fig. 5 shows how to approximate a classical Brownian motion; also in this case the attractor works better than the Eulero scheme for simulation (for details, see [Ia06]).

References

[Ba85] Barnsley, M.F., Ervin, V., Hardin, D., Lancaster, J.: Solution of an inverse problem for fractals and other sets. Proc. Natl. Acad. Sci., **83**, 1975–1977 (1985)

[Ba89] Barnsley, M.F.: Fractals Everywhere. Academic Press, New York (1989)

[Ba97] Barnsley, M.F., Demko, S.: Iterated function systems and the global construction of fractals. Proc. Roy. Soc. London Ser. A, **399**, 243–275 (1985)

[Ca92] Cabrelli, C.A., Forte, B., Molter, U.M., Vrscay, E.R.: Iterated fuzzy set systems: a new approach to the inverse problem for fractals and other sets. J. Math. Anal. Appl., **171**, 79–100 (1992)

[Fo99] Forte, B., Mendivil, F., Vrscay, E.R.: IFS operators on integral transforms. In: Fractals: Theory and Applications in Engineering. Springer Verlag, 81–101 (1999)

[Fo98] Forte, B., Vrscay, E.R.: IFS operators on integral transforms: Theory of generalized fractal transforms. In: Y.Fisher (ed.) Fractal Image Encoding and Analysis, NATO ASI Series F. Springer Verlag, 81-101 (1998)

[Fo981] Forte, B., Vrscay, E.R.: Inverse problem methods for generalized fractal transforms. In: Y.Fisher (ed.) Fractal Image Encoding and Analysis, NATO ASI Series F. ibid

[Hu81] Hutchinson, J.: Fractals and self-similarity. Indiana Univ. J. Math., **730**, 713-747 (1981)

[Hu07] Hutchinson, J., La Torre, D., Mendivil, F.: IFS Markov operators on multimeasures. Preprint (2007)

[Ku07] Kunze, H., La Torre, D., Vrscay, E.R.: Contractive multifunctions, fixed point inclusions and iterated multifunction systems. J. Math. Anal. Appl., **330**, 159–173 (2007)

[Ia05] Iacus, S.M., La Torre, D.: Approximating distribution functions by iterated function systems. J. Appl. Math. Dec. Sc., **1**, 334–345 (2005)

[Ia051] Iacus, S.M., La Torre, D.: A comparative simulation study on the IFS distribution function estimator. Nonlin. Anal.: Real World Appl., **6**, 774–785 (2005)

[Ia06] Iacus, S.M., La Torre, D.: IFSM Representation of Brownian Motion with Applications to Simulation. In: Aletti and oth. (Eds) Math Everywhere – Deterministic and Stochastic Modelling in Biomedicine, Economics and Industry. Springer, 115–124 (2006)

[La06] La Torre, D., Mendivil, F., Vrscay, E.R.: IFS on multifunctions. In: Aletti and oth. (Eds) Math Everywhere – Deterministic and Stochastic Modelling in Biomedicine, Economics and Industry. Springer, 125–135 (2006)

[La061] La Torre, D., Vrscay, E.R., Ebrahimi, M., Barnsley, M.F.: A method of fractal coding for measure-valued images. Preprint (2006)

[Me97] Mendivil, F., Vrscay, E.R.: Correspondence between fractal-wavelet transforms and iterated function systems with grey-level maps. In: J. Levy-Vehel, E. Lutton and C. Tricot (eds.) Fractals in Engineering: From Theory to Industrial Applications. Springer Verlag, London, 125–137 (1997)

[Me02] Mendivil, F., Vrscay, E.R.: Fractal vector measures and vector calculus on planar fractal domains. Chaos, Solitons and Fractals, **14**, 1239-1254 (2002)

[Vr99] Vrscay, E.R.: A generalized class of fractal-wavelet transforms for image representation and compression. Can. J. Elect. Comp. Eng., **23**, 69-84 (1998)

Remarks on Insured Loan Valuations

Mariarosaria Coppola, Valeria D'Amato and Marilena Sibillo

Summary. The paper concerns the case of the insured loan based on an amortization schedule at variable interest rates, hooked at opportune rate indexes. Using the cash flow structure as a basis, the aim is the evaluation of the mathematical provision of a portfolio in a fair value approach. In this environment, the complexity of the life insurance contract market leads to practical valuation management focused on the choice of the most suitable mortality table and discounting process. A numerical application is proposed, for comparing the reserve fair values referred to two insured loans based on an amortization schedule, the first calculated at a fixed rate and the second, the alternative offered in the market, at a variable rate.

Key words: Fair value; Insured loan; Amortization schedule; Cox-Ingersoll-Ross model; Lee-Carter survival probabilities.
JEL Classification: G22, G28, G13

1 Introduction

The International Boards working in the life insurance business accounting field are defining a proper risk assessment in the solvency tool. The pursued aim of putting into effect at the same time the homogenization and the correct information about the business of the different insurance companies, leads the reserve quantification to a mark-to-market valuation of the insurance liabilities, in other words the *fair value*. The *financial risk*, referred to the uncertainty due to movements of the evaluation interest rates and the *demographic risk*, represented by the uncertainty both accidental and systematic in the insured future lifetimes, are the two main risk sources affecting such kind of portfolio evaluations. The paper concerns the insured loan, an insurance product strictly connected to the common operation of a loan repaid by the financial amortization method. The aim of the contract is to square the debt in the case of the insured-borrower predecease.

The loan insurance industry is going through a period of deep evolution, framed in the general increasing attention for guarantee instruments. In case of the borrower/insured's predecease, it can often happen that the banks are interested in providing or requesting the insured loan, and/or the borrower himself wants to set the heirs free

Cira Perna, Marilena Sibillo (Eds.): Mathematical and Statistical Methods in Insurance and Finance

from his obligations. By ensuring that the loan is paid off in the case of the borrower's death, we can say that the insured loan contract, underwritten by the person asking for a loan, is designed to give security to both the contractors.

From the strictly actuarial point of view, this is the framework of an n-year term life insurance with decreasing insured sums.

Different kinds of loans are offered in the market, and most of them are repaid at variable interest rates hooked at a rate index precisely defined in the contract.

In these basic contractual hypotheses, the paper presents a model for the fair valuation of the mathematical provision of an insured loan portfolio, taking into account the two risk sources introduced above. As in [CDLS04, CDLS05 and CDAS06], the approach we follow is the fair value calculation at current values, using the best estimate of the current interest and mortality rates. An application is presented, framed in the particular case of two loans repayable at a fixed or variable interest rate proposed as "financially equivalent" in the market. The aim of the numerical example is to analyze the different fair value behaviour, when the two equivalent amortization schedules can be considered as fixed by the market itself at the evaluation time.

2 The Insured Loan Portfolio: Cash Flow Structure and Reserve Fair Value

On the basis of a general amortization method at variable interest rates (described in Section 3 in more detail), the insured loan contract with duration n binds the insurer to repay the lender the obligations still due by the borrower, if one dies during the contract duration; at time h ($h = 1, 2, \ldots n$), they consist of the outstanding balance (the residual debt) D_{h-1} resulting at time $h-1$ plus the interest on this sum for the period *h-1,h* (cf. [CDAS06]). The value of the benefit payable at time h if the insured-borrower aged x at issue dies during the h-th period and the probability of this event are respectively:

$$B_h = D_{h-1}(1 + i^*_{h-1,h}) \text{ and } {}_{h-1/h}q_x\,, \tag{1}$$

where $i^*_{h-1,h}$ represents the interest rate applicable in the considered period.

Let us suppose that an insured loan with duration n is issued on an insured aged x, with premiums payable at the beginning of each period till the insured is alive or up to m payments ($1 \le m \le n$), and benefit payable at the end of the period of the insured's death, if this event occurs before n.

Considering that the benefit is the sum of the outstanding balance at the beginning of the period and the periodic interest due to that sum and indicating by k_x the curtate future lifetime of the insured, within a deterministic scenario and in the case of anticipated premium payments, the flow at time h is given by the following scheme:

$$X_h = \begin{cases} -{}_{/m}P_{x,h+1} & 1 \le h \le m-1 & k_x > h \\ 0 & m \le h \le n & k_x > h \\ D_{h-1}(1 + i^*_{h-1,h}) & 1 \le h \le n & h-1 \le k_x \le h\,, \end{cases} \tag{2}$$

where $X_0 = -_{/m}P_{x,1}$ and $-_{/m}P_{x,h+1}$ is the $(h+1)$-th premium payable at the beginning of the h-th year.

The above scheme relating to a single contract can be extended, still in a deterministic approach, to a portfolio of c homogeneous insured loans, each one with duration n and issued on a borrower aged x.

Indicating by n_h the number of survivors at time h, the generic cash flow connected to the entire portfolio is given by:

$$\begin{aligned} f_0 &= -c_{/m}P_{x,1} && h = 0 \\ f_h &= -_{/m}P_{x,h+1}n_h + D_{h-1}(1+i^*_{h-1,h})(n_{h-1} - n_h) && h = 1, 2, ..., m-1 \\ f_h &= D_{h-1}(1+i^*_{h-1,h})(n_{h-1} - n_h) && h = m, m+1, ..., n\,. \end{aligned}$$

It can be seen that the schemes can be easily fit to the cases in which periodic premiums and benefits are not due in the same instant and to the common case of a single premium, and of course to the case of an amortization scheduled at a fixed interest rate.

Introducing the stochastic hypotheses for mortality and financial events, let us consider the probability space $\{\Omega, \Im, \wp\}$ originated from the two probability spaces $\{\Omega, \Im', \wp'\}$ and $\{\Omega, \Im'', \wp''\}$, referred respectively to the financial and the demographic information flow (cf. [CDLS05]). The filtration in t of $\Im$ is $\Im_t \subseteq \Im$, with $\Im_t = \Im'_t \cup \Im''_t$ and $\{\Im'_t\} \subseteq \Im', \{\Im''_t\} \subseteq \Im''$.
Under the usual hypotheses of the competitive market (cf. [BH06]), we indicate by:

- $\widetilde{N}_h$, the random variable representing the number of survivors at time h belonging to the group of those, among the c initial insureds at time 0, are living at time t, with $t < h$.
- $v(t, h)$, the stochastic present value at time t of one monetary unit due at time h.
- L_t, the stochastic loss in t $(t < h)$ of the portfolio of the initial c contracts.
- F_t, the stochastic flow at time t.
- $K_{x,t}$, the random curtate future lifetime at time t of the insured aged x at issue.

Considering that the fair value of the mathematical provision is the net present value of the residual debt toward the policyholders, calculated at current interest rates and at current mortality rates, we obtain the following formula, in which F_h indicates the stochastic flow referred to the time h:

$$\begin{aligned} E[L_t/\Im_t] &= E\Big[\textstyle\sum_{h=t+1}^{n}[\widetilde{N}_{h\,/m}P_{x,h+1} \\ &\quad + D_{h-1}(1+i^*_{h-1,h})](\widetilde{N}_{h-1} - \widetilde{N}_h)v(t,h)/\Im_t\Big] \\ &= E\left[\textstyle\sum_{h=t+1}^{n}(F_h v(t,h)/\Im_t)\right]. \end{aligned} \tag{3}$$

Equation (3) is calculated according to a risk neutral valuation, in which the stochastic loss at time t in its fair value form replicates the flow F_h at time h by a trading strategy of units of Zero Coupon Bonds issued at time t and maturing at time h.

The expected value (3) is considered under the hypotheses of the existence of an opportune risk neutral probability measure in a complete market. The question of the incompleteness of the insurance product market, due to the demographic component, is a crucial question in fair value insurance product assessment; the tool is much discussed in the scientific research environment and is fronted in several papers. For example, in [BH06] the question is solved by proposing a risk neutral probability expected value obtained using the most suitable probability measure for the demographic component, setting the market complete in relation to both the systematic and unsystematic mortality risks.

In this general statement, we will proceed in the calculations supposing that the variables $K_{x,t}$ are independent and identically distributed, the random variables F_h and $v(t, h)$ are independent and identically distributed conditioning on the interest process, and the two risk sources $K_{x,t}$ and $v(t, h)$ are independent. Equation (3) can be expressed as follows:

$$E[L_t/\Im_t] = c_t\, p_x \left\{\left[{}_{/m}P_{x,h+1}\; E\left(v(t,h)/\Im_t\right) {}_h p_{x+t} \right.\right.\\ \left.\left. + {}_{h-1/h}q_{x+t} D_{h-1} E\left((1+i^*_{h-1,h})v(t,h)/\Im_t\right)\right]\right\}. \tag{4}$$

3 The Application to a Case of Equivalent Products

Introduction

The aim of this section is to compare the different behaviour of the mathematical reserve fair value of two products offered as alternatives in the loan market at the time of valuation $t = 0$, characterized by an amortization schedule at a fixed rate in the first case and at a variable rate in the second. We will consider the circumstances in which they are proposed alternatively as the realization of the financial equivalence condition of the two products at time 0, so that, for the sake of this application, the two amortization schedules can be considered as defined by the market itself. In this order of ideas, we will consider both the amortization schedules as fixed at the time of valuation; from the financial risk point of view, this means we will focus our attention on the volatility of the valuation interest rates.

The two cases are referred to a portfolio of $c = 1000$ homogeneous insured loan policies, issued to a unitary capital, repayable in $n = 20$ half-years. As already stated, at the end of the period in which the borrower/insured dies, if this event happens before n, the insurer pays to the lender the outstanding balance existing at the beginning of the period plus the (fixed or variable) interest on this sum for the same period, in this way extinguishing the borrower's obligations. The amortization schemes that the market expresses as equivalent financial operations are at the fixed rate 4.25% and at a variable rate linked to the half-yearly Euribor rate as reference index. The payments are due at the end of 20 half-years beginning from $t = 0$. As commonly used, the two cases count the payment of a single premium paid at the contract issue time.

The numerical procedure for the Euribor rate estimation

Estimation of the future Euribor interest rates (as well as the evolution in time we will consider next, referring to fair evaluation rates) suffers the long duration of these scenarios. The necessity of quantifying imposes the choice of the best estimation tool and to this aim several methods can be applied, such as items in the historical series analysis fields or information taken in the money market, using available prices of cash today's Libor rates, future and interest rate swap with respect to the different duration of the forecasting.

In this paper, in order to give an example, we have analyzed the Euribor rate historical series from January 1, 2006 to September 22, 2006, data available on `www.euribor.org`, website controlled by the *FBE-European Banking Federation*. The rates refer to 0, 1, 2, 3, 4, 5, 6 months (0 means a very short term, in our case 1 week) and are daily registered. Supposing the loan issued on January 2, 2006, following [Ben01] we work on cross sections successively calculated over the data related to each half-year and proceed with a 3rd degree polynomial regression over them; in this way we begin to obtain the first regression equation for the first data set and consequently calculating the estimation of the oscillations the data set will undergo in the next period. This method furnishes a good approximation of the term structure of the considered rates, as is asserted in [Ben01]. For example, the following are the cross section rates related to June 1, 2006 over which the polynomial regression is done:

Table 1. The cross section rates related to June 1, 2006.

months	0	1	2	3	4	5	6
rates	2.621%	2.821%	2.901%	2.944%	3.022%	3.084%	3.126%

This recursive method, followed for each half-year, allows the rate referred till the last half-year to be obtained, the second one of 2016.

In Table 2 the Euribor estimated rates for the half-years considered in the example are reported.

Table 2. The Euribor estimated rates for the half-years.

Half years	Rates	Half years	Rates
1	2.64%	11	5.07%
2	3.13%	12	5.35%
3	3.16%	13	5.63%
4	3.43%	14	5.91%
5	3.71%	15	6.19%
6	3.99%	16	6.47%
7	4.27%	17	6.75%
8	4.23%	18	7.03%
9	4.51%	19	7.31%
10	4.79%	20	7.59%

The amortization scheme at variable interest rates

The refund is settled starting from a theoretical amortization schedule with constant periodic payment amounts, obtained using an opportune interest rate: as in [VDP93], we fix this rate in the average of the estimated half-yearly Euribor rates. This theoretical amortization scheme furnishes the principal repaid amounts the borrower will correspond to the lender at the end of the future n periods. Besides these undertakings, the borrower has to correspond the interest paid amounts, calculated each period at the periodical Euribor interest rate existing in the market at the beginning of the period itself. On the contrary of the theoretical scheme, in the effective one, the entire periodical payment amounts are variable, their value being connected to the Euribor rate trend.

Fair valuation hypotheses

On the basis of the amortization scheme calculated in according with what is above described, the evaluation of the reserve fair value follows by means of (3). The evolution in time of the evaluation stochastic rates usable for the financial instruments constituting the Zero Coupon Bond replicating portfolio, is described by the Cox-Ingersoll- Ross square root model, represented by the following SDE: $dr_t = -k(r_t - \gamma)dt + \sigma\sqrt{r_t}dB_t$ in which k and σ are positive constants, g is the long term mean and B_t a Brownian motion. As in [CDFDLS05], we assign the following values to the parameters: $r_0 = 0.0172$, $g = 0.0452$, $s = 0.0052$.

The survival probabilities used in the application are deduced by the Lee-Carter model. This law is considered as a good description of the survival phenomenon, being able to correct itself year by year, capturing the changes in the trend due in particular to the longevity phenomenon. Moreover, as shown in [DF05], this model furnishes a very good representation of the survival phenomenon in intervals of 8-10 years and so is particularly appropriate in the case of the considered portfolio.

The probabilities have been obtained by means of the tables of the parameters referred to the considered period, as calculated in [CDFDLS05].

Results

In Fig. 1 the comparison between the reserve fair value of the insured loan portfolio in the two cases of the above amortization schemes at fixed and variable interest rates are reported. The example is referred to the fixed rate 4.25%, alternatively offered in the market for loan of 20 half-year periods with interest rates variable with the Euribor rates. The graphical result shows the decisive influence of the relationship between the Euribor term structure and the fixed rate. As can be seen from Table 1, after the 3rd year, the fixed rate starts to be less than the Euribor rates: this component appears to be orienting in determining the two different behaviours and the intersection between them.

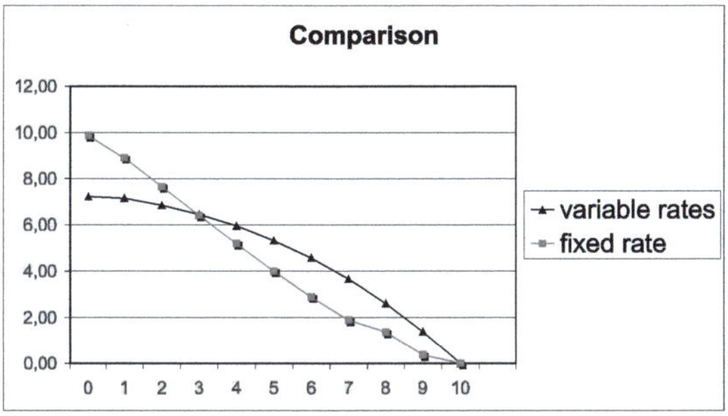

Fig. 1. A comparison between the reserve fair value in the two cases of amortization schemes at fixed and variable interest rates.

4 Conclusions

In this paper, the fair valuation of the reserve in the case of a portfolio of insured loans is presented, considering the case of a basic amortization scheme scheduled at variable interest rates. A practical example of application is reported, with the aim of comparing the different behaviour of the current values of the portfolio reserve when the amortization scheme is at fixed or variable interest rate. The application refers to products offered in the market as alternative between them and the reserve is valued at the moment in which the contracts are issued. As the two schemes proposed are "equivalent", the aim is to refer the comparison to products financially equivalent at the time of valuation and as a consequence the two amortization schemes are considered entirely fixed in $t = 0$. Further development of the item will be developed in two directions: the reserve fair valuation considered in a stochastic framework for the amortization interest repaid and the extension of the same approach to other fields, as for example to the cases in which the insurance contract is referred to the borrower default probabilities, linking the insurer obligations to events not referred exclusively to human life.

References

[BH06] Ballotta, L., Haberman, S.: The fair valuation problem of guaranteed annuity options: The stochastic mortality environment case. Insurance: Mathematics and Economics, **38**, 195–214 (2006)

[Ben01] Benninga, S.: Modelli Finanziari. McGraw Hill, Italia (2001)

[CDLS05] Coppola, M., Di Lorenzo, E., Sibillo, M.: Fair valuation scheme for life annuity contracts. Proceedings of the XI "International Symposium on Applied Stochastic Models and Data Analysis" (ASMDA 2005). ISBN 2- 908849-15-1, 893–901 (2005)

[CDAS06] Coppola, M., D'Amato,V., Sibillo, M.: Fair value and demographic aspects of the insured loans. 10th International Congress on Insurance: Mathematics and Economics. Leuven (Belgium) (2005)

[CDLS04] Cocozza, R., Di Lorenzo, E., Sibillo, M.: Life insurance risk indicators: a balance sheet approach.Proceedings of the 9th International Congress on Insurance: Mathematics and Economics. Rome (2004)
[CDFDLS05] Cocozza, R., De Feo, D., Di Lorenzo, E., Sibillo, M.: On the financial risk factors in fir valuation of the mathematical provision. Proceedings of Astin Conference, Zurich (2005)
[DF05] De Feo, D.: An empirical method for human mortality forecasting. An application to Italian data. Problems and Perspectives in Management, **4**, 201–219, (2005)
[J04] Jorgensen, P. L.: On accounting standards and fair valuation of life insurance and pension liabilities. Scandinavian Actuarial Journal, **5**, 372–394 (2004)
[VDP93] Volpe Di Prignano, E.:Manuale di Matematuica Finanziaria. II ed. Edizioni ESI, Napoli (1993)

Exploring the Copula Approach for the Analysis of Financial Durations

Giovanni De Luca, Giorgia Rivieccio and Paola Zuccolotto

Summary. The object of the paper is to compare of two approaches for the analysis of financial durations. The first is the parametric approach (Autoregressive Conditional Duration model) implemented using the exponential, the Weibull, the Burr and the Pareto density functions. The second makes use of bivariate and trivariate copula functions.

Key words: Financial duration; Autoregressive conditional duration; Copula function.

1 Introduction

The seminal work of [ER98] has opened the interest in ultra-high frequency financial data, also known as tick-by-tick data. They are irregularly spaced time series which enable the researcher to investigate the process of transactions of a financial asset traded on a financial market. The Autoregressive Conditional Duration (ACD) model is now a consolidated statistical tool to describe the time between the occurrence of two market events.

In their pioneering work, [ER98] made use of exponential and Weibull random variable for the durations. Many authors tried to extend the analysis relying on more elaborated densities, such as the Pareto or the Burr.

In this paper we would like to compare the traditional parametric approach with a semi-parametric approach based on the copula function. In particular, we will make comparisons using bivariate and trivariate copulas.

The paper is structured as follows. In Section 2 the ACD framework is presented. Section 3 briefly reviews the concept of copula function. In Section 4 data analysis is carried out, and Section 5 concludes.

2 ACD Models

The class of ACD models is aimed at modelling the durations between two market events, such as price changes or bid-ask spreads. Let X_i be the duration between two

Cira Perna, Marilena Sibillo (Eds.): Mathematical and Statistical Methods in Insurance and Finance

observations at times t_{i-1} and t_i. [ER98] proposed the ACD(q, p) model

$$\begin{aligned} X_i &= \phi(t_i)\Psi_i\epsilon_i \\ \Psi_i &= \omega + \sum_{j=1}^{q} \alpha_j x_{i-j} + \sum_{j=1}^{p} \beta_j \Psi_{i-j}, \end{aligned} \tag{1}$$

where $\phi(t_i)$ is a deterministic daily seasonal component and $x_i = X_i/\phi(t_i)$ is the seasonally adjusted duration at time t_i. In addition $\omega > 0$, $\alpha_j, \beta_j \geq 0$ and $\sum_j \alpha_j + \sum_j \beta_j < 1$ in order to ensure, respectively, the positivity of the conditional expected duration and the stationarity of the process.

Assuming ϵ_i identically and independently distributed with $E(\epsilon_i) = 1$, it is easy to see that $E(x_i|\mathcal{F}_{i-1}) = \Psi_i$, where $\mathcal{F}_{i-1}$ is the information at time t_{i-1}. As a result, Ψ_i can be interpreted as the i-th expected (deseasonalized) duration conditionally on the information at time t_{i-1}.

When $q = p = 1$ the popular ACD(1,1) model is obtained. It is a parsimonious model which adequately fits durations in most cases. Model (1) becomes

$$\begin{aligned} x_i &= \Psi_i\epsilon_i \\ \Psi_i &= \omega + \alpha x_{i-1} + \beta\Psi_{i-1}. \end{aligned}$$

In order to estimate the model using parametric methods, a distributional assumption on ϵ_i is needed. The traditional hypotheses, the exponential and the Weibull (EACD and WACD, [ER98]), rely on the theory of point processes, supporting the whole apparatus of ACD models. Nonetheless, in the literature more refined proposals are present such as, for example, the Pareto (PACD, [DZ03]) and the Burr (BACD, [GM00]) which can be viewed, respectively, as infinite mixtures of exponential and Weibull distributions. The use of mixtures of distributions is thought to take into account the heterogeneity of traders in the market ([DZ03]) and some recent contributions propose models allowing for a regime-switching behavior of the mixing distribution ([DZ06]).

Given a vector x of T observed durations, the parameters of the ACD model and those characterizing the distribution can be collected in the vector η and jointly estimated via ML using the following log-likelihood functions, respectively for the EACD, WACD, PACD, BACD model:

$$l(\eta; x) = -\sum_{i=1}^{T} \left\{ \log \Psi_i + \frac{x_i}{\Psi_i} \right\},$$

$$l(\eta; x) = \sum_{i=1}^{T} \left\{ \log \frac{\gamma}{x_i} + \gamma \left[\log \frac{x_i \Gamma\left(1 + \frac{1}{\gamma}\right)}{\Psi_i} \right] - \left[\frac{x_i \Gamma\left(1 + \frac{1}{\gamma}\right)}{\Psi_i} \right]^{\gamma} \right\},$$

$$l(\eta; x) = \sum_{i=1}^{T} \left[(\theta + 1) \log \theta_1 + \log \frac{\theta + 1}{\Psi_i} - \log \left(\theta + \frac{x_i}{\Psi_i} \right)^{\theta+2} \right]$$

and

$$l(\eta; x) = \sum_{i=1}^{T} \left\{ \log \kappa - \log(c\Psi_i)^{\kappa} + \log x_i^{\kappa-1} - \log \left[1 + \sigma^2 \left(\frac{x_i}{c\Psi_i} \right)^{\kappa} \right]^{\frac{1}{\sigma^2}+1} \right\},$$

where

$$c = \frac{\left(\sigma^2\right)^{1+\frac{1}{\kappa}} \Gamma\left(1 + \frac{1}{\sigma^2}\right)}{\Gamma\left(1 + \frac{1}{\kappa}\right) \Gamma\left(\frac{1}{\sigma^2} - \frac{1}{\kappa}\right)}.$$

3 Copula Functions

A copula is a multivariate distribution function H of random variables $X_1, \ldots, X_n$ with standard uniform marginal distributions $F_1, \ldots, F_n$ defined on the unit n-cube $[0, 1]^n$ with the following properties:

1. The range of the copula $C(u_1, \ldots, u_n)$ is the unit interval $[0, 1]$.
2. $C(u_1, \ldots, u_n) = 0$ if any $u_i = 0$ for $i = 1, 2, \ldots, n$.
3. $C(1, \ldots, 1, u_i, 1, \ldots, 1) = u_i$ for $i = 1, 2, \ldots, n$ and for all $u_i \in [0, 1]$.

$F_i = P(X_i \leq x_i) = u_i$ is uniform in $[0, 1]$ for all $i = 1, 2, \ldots, n$.

One of the most important copula based theorem is the Sklar's theorem, which justifies the role of copula as a dependence function. Let H be a joint distribution function with marginal distribution functions $F_1, \ldots, F_n$, then a copula function C exists such that:

$$C(u_1, \ldots, u_n) = H(x_1, \ldots, x_n).$$

Therefore, the joint distribution is split into two components: the unconditional marginal distributions and the dependence structure given by the copula. The copula tool allows the whole dependence structure that characterizes the relationships among the variables to be described. For this reason, the copula function has recently become a very significant quantitative technique to handle many financial time series analyses characterized by a remarkable temporal dependence.

The advantage of the copula method consists of modelling the duration process with a semiparametric approach where the copula is parameterized and controls the dependence structure, while the unconditional marginal distributions are left unspecified so that it is possible to choose between all kinds of marginal distributions ([Jo97 and Ne99]).

In the literature there are two main families of copulas, the elliptical copulas which are copula functions of elliptical distributions and the Archimedean ones, based on the definition of a generator function $\Phi(t) : t \in R^+$, continuous, decreasing, convex and such that $\Phi(1) = 0$. The Archimedean copula function is defined as

$$C(u_1, \ldots, u_n) = \Phi^{-1}\left(\Phi(u_1) + \ldots + \Phi(u_n)\right),$$

with $\Phi^{-1}(\cdot)$ completely monotonic on $[0, \infty]$. For a detailed description, see [Ne99].

In this work, we have adopted two Archimedean copulas: the Clayton and the Frank.

For the Clayton copula, the generator $\Phi_\alpha(t)$ is given by $\left(t^{-\alpha} - 1\right)$ with $\alpha \in [-1, 0) \cup (0, +\infty)$ when $n = 2$ and strictly positive for $n \geq 3$. The copula function is

$$C_C(u_1, \ldots, u_n) = \left(\sum_{i=1}^{n} u_i^{-\alpha} - n + 1\right)^{-1/\alpha}.$$

The generator of the Frank copula is

$$\Phi_\alpha(t) = -\log \frac{\exp(-\alpha t) - 1}{\exp(-\alpha) - 1},$$

and the expression of the copula function is

$$C_F(u_1, \ldots, u_n) = -\frac{1}{\alpha} \log\left(1 + \frac{\prod_{i=1}^{n} \left(\exp(-\alpha u_i) - 1\right)}{\left(\exp(-\alpha) - 1\right)^{n-1}}\right).$$

The range of the parameter α is equal to $(-\infty, 0) \cup (0, +\infty)$ in the bivariate case and strictly positive in the multivariate case.

The estimation of the parameter α can be carried out through the canonical maximum likelihood method, a two-steps procedure without assumptions on the parametric form for the marginals. In the first, the sample data $(x_{1i}, \ldots, x_{ni})_{i=1}^{T}$ are transformed into uniform variates $(\hat{u}_{1i}, \ldots, \hat{u}_{ni})_{i=1}^{T}$ using the empirical distribution functions as estimates of the marginal distribution functions. In the second, the copula parameter α is evaluated via maximum likelihood,

$$\hat{\alpha} = \arg\max \sum_{i=1}^{T} \log c(\hat{u}_{1i}, \ldots, \hat{u}_{ni}; \alpha),$$

where $c(\hat{u}_{1i}, \ldots, \hat{u}_{ni}; \alpha)$ is the copula density, that is n-th derivative of the copula function with respect to the marginals, depending on the parameter α.

4 Data Analysis

The empirical analysis has focused on duration data of the transactions involving a price change in the Italian stock Comit in the month of February 2000. The total number of obervations is 8221. We removed the daily seasonal component after estimating it using a cubic spline with nodes set at each hour.

Table 1. Estimate (standard error) of the exponential and Weibull ACD models.

Parameter	Exponential	Weibull
ω	0.0019 (0.0007)	0.0018 (0.0007)
α	0.0317 (0.0036)	0.0317 (0.0037)
β	0.9667 (0.0038)	0.9667 (0.0040)
γ	–	0.9576 (0.0081)
Diagnostics on residuals		
LB(20) (p-value)	0.1071	0.1065
Residuals mean	1.0003	1.0012
Mean of the assumed distribution	*1*	*1*
Residuals variance	1.2321	1.2346
Variance of the assumed distribution	*1*	*1.0911*

Table 2. Estimate (standard error) of the Pareto and Burr ACD models.

Parameter	Pareto	Burr
ω	0.0016 (0.0007)	0.0016 (0.0007)
α	0.0321 (0.0039)	0.0322 (0.0039)
β	0.9665 (0.0041)	0.9666 (0.0041)
θ	9.7711 (1.4021)	–
κ	–	1.0457 (0.0152)
σ^2	–	0.1443 (0.0218)
Diagnostics on residuals		
LB(20) (p-value)	0.1140	0.1157
Residuals mean	1.0001	0.9990
Mean of the assumed distribution	*1*	*1*
Residuals variance	1.2330	1.2305
Variance of the assumed distribution	*1.2280*	*1.2640*

For the parametric approach, we estimated by maximum likelihood the EACD, WACD, PACD, BACD models ($p = q = 1$), that is under the assumptions of exponential, Weibull, Pareto and Burr distributed innovations (Tables 1 and 2). As expected, the estimates of the ACD parameters do not significantly differ from one model to the other, and the Ljung-Box (LB) statistics for the hypothesis of uncorrelated residuals are always satisfactory. However, the parameters of the Pareto and the Burr distributions are highly significant, which is the first evidence recommending the use of PACD and BACD models. In addition, the averages of the residuals $\hat{\epsilon}_i = x_i/\widehat{\Psi_i}$ are always strictly close to one, but the variance is far from unity and more coherent with the mixture of distribution models (PACD and BACD). The parametric approach seems to ensure a good fit to data, expecially when more refined distributional assumptions are used as will be clearly confirmed later, with the analysis of the density forecasts.

On the other hand, the semiparametric approach (see [SN05]) was carried out using the Clayton and Frank copulas for the dependence between x_i and x_{i-1} ($n = 2$) and then among x_i, x_{i-1} and x_{i-2} ($n = 3$). The estimates of the models are shown in Table 3.

In order to compare the two approaches we use the density forecast evaluation as outlined in [BG04]. The probability integral transforms of the one-step-ahead forecasts of the durations are independently and uniformly distributed in [0, 1] if the model is correctly specified.

For an n-dimensional copula function, the probability integral transform is given by the conditional copula, obtained as

$$C(u_n|u_1,\ldots,u_{n-1}) = \frac{\partial^{n-1}C(u_1,\ldots,u_n)/\partial u_1\ldots\partial u_{n-1}}{\partial^{n-1}C(u_1,\ldots,u_{n-1})/\partial u_1\ldots\partial u_{n-1}}.$$

The eight histograms of the in-sample one-step-ahead forecasts are reported in Figs. 1 and 2. In particular the copula functions appear to have a better performance. The Kolmogorov-Smironv statistics for the hypothesis of uniformity (Table 4) confirm the graphical evidence.

Table 3. Estimate (standard error) of the copula functions.

Model	Parameter α
Biv. Clayton	0.1325 (0.0141)
Biv. Frank	0.8582 (0.0667)
Triv. Clayton	0.1217 (0.0088)
Triv. Frank	0.7400 (0.0419)

Table 4. Kolmogorov-Smirnov statistics.

Model	KS statistics	Model	KS statistics
Exponential	0.028	Biv. Clayton	0.006
Weibull	0.016	Biv. Frank	0.005
Pareto	0.009	Triv. Clayton	0.006
Burr	0.009	Triv. Frank	0.006

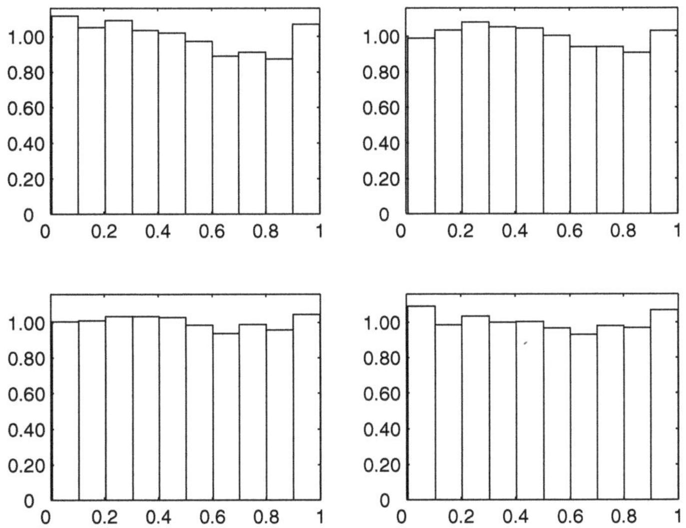

Fig. 1. Histogram of one-step-ahead forecasts for exponential (*top left*), Weibull (*top right*), Pareto (*bottom left*) and Burr (*bottom right*).

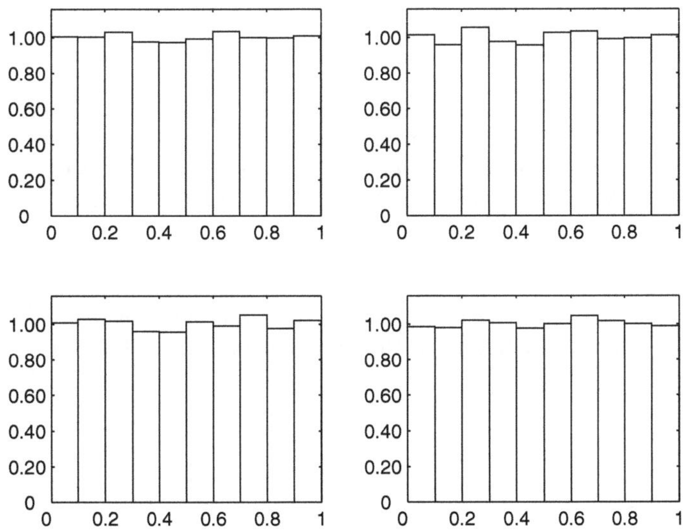

Fig. 2. Histogram of one-step-ahead forecasts for bivariate Clayton copula (*top left*), bivariate Frank copula (*top right*), trivariate Clayton copula (*bottom left*) and trivariate Frank copula (*bottom right*).

5 Concluding Remarks

In this paper two different methods for the modelling of financial durations deriving from tick-by-tick datasets are compared. On one hand we examined a traditional, parametric approach based on ACD models and on the other hand a semiparametric one based on the copula function. The results of an application to real data showed some interesting points. For the analyzed dataset, both the parametric and the semiparametric approach provide good in-sample results, but the copula functions exhibit a small superiority. Secondly, while in the context of ACD models there is no doubt about the (well known) supremacy of BACD and PACD models on the traditional EACD and WACD, it is not clear if the Clayton should be preferred to the Frank copula or if the bivariate approach performs better than the trivariate one. A deeper investigation into this is encouraged, in order to verify both the out-of-sample performance of the two approaches and the differences among the copula specifications.

References

[BG04] Bauwens, L., Giot, P., Grammig, J., Veredas, D.: A comparison of financial durations models via density forecast. International Journal of Forecasting, **20**, 589–609 (2004)

[DZ03] De Luca, G., Zuccolotto, P.: Finite and infinite mixture for financial duration. Metron, **61**, 431–455 (2003)

[DZ06] De Luca, G., Zuccolotto, P.: Regime-switching Pareto distributions for ACD models. Computational Statistics and Data Analysis, **51**, 2179–2191 (2006)

[ER98] Engle, R. F., Russell, J.R.: Autoregressive Conditional Duration: a new model for irregularly spaced transaction data. Econometrica, **66**, 1127–1162 (1998)

[GM00] Grammig, J., Maurer, K.: Non-monotonic hazard functions and the autoregressive conditional duration model. Econometrics Journal, **3**, 16–38 (2000)

[Jo97] Joe, H.: Multivariate models and dependence concepts. Chapman & Hall, London (1997)

[Ne99] Nelsen, R. B.: An Introduction to Copulas, Lectures Notes in Statistics. Springer, New York (1999)

[SN05] Savu, C., Ng, W. L.: The SCoD Model. Analyzing durations with a semiparametric copula approach. International Review of Finance, **5**, 55–74 (2005)

Analysis of Economic Fluctuations: A Contribution from Chaos Theory*

Marisa Faggini

Summary. The nature of business cycles is a central and conflicting question in macroeconomics. We would like to stress, however, the importance of chaos, in the context of business cycle theories. In fact, those who believe in i.i.d. disturbances simply state that fluctuations are determined by exogenous factors. Chaos supporters, on the other hand, disagree with a linear world and believe that the source of fluctuations is endogenous to the economic system. The aim of paper is to highlight the power of chaos theory to analyze business cycles.

Key words: P10; H3; E6.

1 Introduction

Most economic indicators have elements of a continuing wave-like movement and are partially erratic, serially correlated and more than one periodicity has been identified in their behaviour in addition to long growth trends (Chen 1988). How to identify these movements is a central and conflicting debate in macroeconomics that arises around two opposite approaches: the exogenous-shocks-equilibrium, the endogenous-cycles-disequilibrium. For exogenous-shocks-equilibrium the fluctuations are deviations from a steady growth path determined by exogenous "shocks" like fiscal and monetary policy changes, and changes in technology. Stochastic exogenous disturbances are superimposed upon (usually linear) deterministic models to produce the stochastic appearance of actual economic time series (Prokhorov 2001). In many cases, however, those approaches fail to provide an economic explanation for the serial correlation of error terms and of the exact meaning of the exogenous shocks. According to endogenous-cycles-disequilibrium, deviations from growth trends are consequences of endogenous shocks coming from imperfections of the market. In this sense, endogenous cycles are represented by deterministic oscillators including harmonic cycle and limit cycle (Samuelson 1939, Hicks 1950 and Goodwin 1951). Nevertheless, the real economic time-series do not show the kind of regularity and symmetry that is predicted by those models. Irregular frequencies

* This research is supported by a grant from Miur-Firb RBAU01B49F/001

Cira Perna, Marilena Sibillo (Eds.): Mathematical and Statistical Methods in Insurance and Finance

and different amplitudes are the real feature of economic fluctuations that do not show clear convergence or steady oscillations. This contrast with reality pushed the economists to use non-linear approaches to analyse business cycle models. They are well suited for examining macroeconomic fluctuations and business cycle research because they are powerful tools able to capture stylized facts observed in many financial and economic time series like asymmetries, jumps and time irreversibility, and to model any oscillating phenomenon. The most exciting feature of non-linear systems is their ability to display chaotic dynamics. Models of deterministic chaos are a subcategory of non-linear models characterized by the conjunction of local instability, giving rise to sensitivity to initial conditions, with global stability that effectively restricts the long-term dynamic to attractor's orbits. For this, chaotic phenomena, notwithstanding their seemingly random time paths, are representations of long-term steady states of the dynamic system. Chaos theory could explain the irregularity of economical fluctuation using simple non-linearities and may become more common as economists increasingly familiarize themselves with non-linear techniques of mathematical and statistical analysis from physics. There are some differences, however, between physics and other hard sciences, in which these techniques are largely used, and economics (Faggini 2006b). In some cases topological and not metric instruments must be used. There are important policy reasons to understand the impact of non-linearities and chaos in social systems. First of all, it is possible to have a more realistic description of economic phenomena, and the control of chaotic systems can actually be easier than the control of linear ones, because it might take only a small push to engender a big change in the system. In other words, small, low-cost policy changes could have a large impact on overall social welfare. Therefore, the aim of the paper is, starting from a description of results of traditional approaches to business cycles, to highlight how the chaos theory could contribute to improve the description of economic phenomena fluctuations and why this approach was discharged in favour of the non-linear stochastic approach.

2 Non-linear Deterministic Systems. Is Economy a Chaotic System?

Medio (1992) defines a deterministic system as "one containing no exogenous stochastic variables." Knowledge of all a system's equations and the links between them will lead to accurate prediction in a system (Kutcha 2004). While the deterministic approach assumption is that random behaviour observed in economics time series is the result of non-stochastic process, the stochastic approach assumption states that the structure of the underlying process, either linear or non-linear, is subjected to random shocks. The importance of classifying a time series as deterministic or stochastic cannot be overemphasized, because for example in finance, it represents the difference between evidence for or against the market efficiency hypothesis (Peters 1991). Deterministic and stochastic systems can be linear and non-linear. A linear deterministic model for business cycle was first proposed by Samuelson (1939), which generated damped or explosive cycles. Non-linearities in deterministic models were introduced in terms of limit cycles to explain the self-sustained wavelike movement in economics (Hick 1950 and Goodwin 1951). Linear stochastic models, in particular the class of ARMA models, have been a practical tool for economic analysis and forecasting, nevertheless this class has a number of serious shortcomings for studying economic fluctuations (Potter 1995). Liner autoregressive or ARMA models are only capable of generating realizations with symmetrical cyclical fluctuations and are not capable of accommodating large shocks, shifting trends and structural changes. The theory leading to linear models of business cycles has to be given up as inadequate so that non-linear models are needed to describe the asymmetric fluctuations. Empirical evidence for non-linearities in economic time series fluctuations is reported in Hsieh 1991, Potter 1995 and Brooks 2001. ARCH processes are non-linear stochastic models that show no autocorrelation, have a mean zero, a constant unconditional variance, and most importantly, they have non-constant variances conditional on the past. These techniques allowed detecting in data dynamics otherwise obscured by systematic noise, time varying volatility, and non-stationary trends opening to the analysis of financial and macroeconomic time series. Bera and Higgins (1993) remarked that "a major contribution of the ARCH literature is the finding that apparent changes in the volatility of economic time series may be predictable and result from a specific type of non-linear dependence rather than exogenous structural changes in variables." The results of research in economic and financial data is the evidence of widespread stochastic non-linearity, even though the main effects seem to be exhibited in the variances of the respective distributions. Nevertheless some researchers[2] (Brock et al. 1991, Frank and Stengos 1988), have indicated that generalized ARCH models still leave some evidence of non-linearities in the data. However, what that non-linearity is and how it should be modelled is still an open question. In contrast to ARCH models, chaos represents the stochastic behavior generated by non-linear deterministic systems (Tsonis 1992 and Medio 1992). In fact, many non-linear deterministic systems exhibit a behavior so irregular that most standard randomness tests are unable to distinguish between them and pure white noise. Such non-linear dynamical models characterized (a) by the intrinsically generated stochasticity and (b) by high sensitivity to initial conditions and parameter values, are the subject matter of chaos theory. The seeming irregularity of business cycles using chaos theory, has been employed in Day R.H., (1994), Dechert W.D., (1996), Goodwin R., (1990), Hommes C.H., (1991), Lorenz H.W., (1993) and Puu T., (1997). Moreover, tests for chaos and non-linearity in data were developed. The use of these techniques to test the presence of chaos in time series shows that data quantity and data quality are crucial in applying them. All researchers applying them filtered their data by either linear or non-linear models (Frank and Stengos 1989, Blank 1991, Cromwell and Labys 1993 and Yang and Brorsen 1992, 1993), in most cases by ARCH-type models, then, conducted the chaos analysis on the residuals. The open question is whether chaotic properties of the process are invariant to such

---

[2] Lorenz (1989), Ashley et al. (1986), Ramsey et al. (1990), Granger (1991), Ramsey and Rothman (1996), Lee et al. (1993), Bollerslev et al. (1990), Barnett et al. (1993), Rothman (1994), and Le Baron (1994)

transformations. The simulations have proven that linear and non-linear filters will distort potential chaotic structures (Chen 1993 and Wei and Leuthold 1998). Current tests used to detect non-linear structure often fail to find evidence of non-linearity in aggregated data, even if the data are generated by a non-linear process and don't distinguish between non-linearity or chaos in the economic data (Barnett 2003). Data quantity and data quality in economics are a significant obstacle to chaotic-economic theory, and one of the main reasons why the literature has not reached a consensus on the existence of chaotic dynamics in data (Faggini 2005, 2006a). The consequence of this is resumed in Granger and Terasvirta (1992): "Deterministic (chaotic) models are of little relevance in economics and so we will consider only stochastic models". The question was amplified by Jaditz and Sayers (1993), who reviewed a wide variety of research to conclude that there was no evidence for chaos, but that was not to deny the indication of non-linear dynamics of some sort. Therefore, in order to facilitate the testing of deterministic chaos and to improve our understanding of modern economies, it is worthwhile developing numerical algorithms that work with moderate data sets and are robust against noise. The topological approach provides the basis for a new way to test time series data for chaotic behaviour (Mindlin et 1990). In fact, it has been successfully applied in sciences to detect chaos in experimental data, and it is particularly applicable to economic and financial data since it works well on relatively small data sets, is robust against noise, and preserves time-ordering information (Faggini 2005, 2006b). This methodology makes it possible to reveal correlation in the data that is not possible to detect in the original time series. It does not require either assumptions on the stationarity of time series or the underlying equations of motions. It seems especially useful for cases in which there is modest data availability and it can efficiently compare to classical approaches for analyzing chaotic data, especially in its ability to detect bifurcation (Faggini 2005, 2006b). Recurrence Analysis is particularly suitable to investigate the economic time series that are characterized by noise and lack of data and are an output of high dimensional systems. Proving that the data is chaotic would prove there is a deterministic underlying system generating the data. The proof would be a giant step toward clarifying the "nature" of the economy. Chaotic non-linear systems can endogenize shocks. If the economy is chaotic, then we can create a complete and closed model. This development would aid significantly short run forecasting and control. Detecting chaos in economic data is the first condition to apply a chaotic control to phenomena that generate them. In fact, chaos theory offers attractive possibilities for control strategies (Faggini 2006a) and this point seems particularly relevant for the insights of economic policies. Using sensitivity to initial conditions to move from given orbits to other orbits of attractors means to choose different behaviour of the systems, that is, different trade-off of economic policy. Moreover, the employment of an instrument of control in terms of resources in order to achieve a specific goal of economic policy will be smaller if compared to the use of traditional techniques of control.

3 Conclusion

The failure to find convincing evidence for chaos in economic time series redirected efforts from modeling non-linearity from conditional mean, chaotic systems, toward conditional variance, ARCH-type models (Prokov 2001). Predicting volatility has become the new challenge of non-linear economics. Nevertheless, some researchers have indicated that generalized ARCH models still leave some evidence of unexplained non-linearities in the data. This could suggest redirecting the time series analysis toward chaotic approach as some problems encountered using chaos tests could be overcome. New techniques for the analysis of short time series like economics, so to distinguish stochastic fluctuations from purely deterministic chaotic data, have been developed. The discovery of deterministic chaos changes our view of erratic behaviour of economic fluctuations. Therefore, the salient feature of applying chaotic control is the strong "energy saving", that is resources, to perform economic policy goals. If the system is non-chaotic, the effect of an input on the output is proportional to the latter. Vice versa, when the system is chaotic, the relationship between input and output is made exponential by the sensitivity to initial conditions. We can obtain a relatively large improvement in system performance by using small controls. Resource saving and choosing from different trade-offs of economic policies (many orbits) could be significant motivations to use chaotic models in the economic analysis. To use these models we need to discover chaos in the data, and we have demonstrated that VRA is a useful tool for doing this (Faggini 2005, 2006).

References

[BH03] Barnett, W. and Y. He: Unsolved Econometric Problems in Nonlinearity, Chaos, and Bifurcation, Department of Economics, Washington University-St. Louis Working Papers Working Paper Working Paper. URL: http://econwpa.wustl.edu:80/eps/mac/papers/0004/0004021.pdf. 12/03 (2003)

[BH93] Bera, A. K. and M. L. Higgins: ARCH Models: Properties, Estimation and Testing, Journal of Economic Surveys, Vol. 7, No. 4, pp. 307–366 (1993)

[B91] Blank S. C.: Chaos in Futures Markets: A Nonlinear Dynamically Analysis, The Journal of Futures Markets, vol. 11, pp. 711–728 (1991)

[BHL91] Brock W. A., D. Hsieh and B. LeBaron: Nonlinear Dynamics, Chaos, and Instability: Statistical Theory and Economic Evidence. Cambridge, Massachusetts: MIT Press (1991)

[B01] Brooks, C.: A double-threshold GARCH model for the French Franc/Deutschamrk exchange rate, Journal of Forecasting, 20, 135–143 (2001)

[C88] Chen, P.: Empirical and Theoretical Evidence of Economic Chaos, System Dynamics Review, vol. 4, nr. 1–2, 81–108. (1988)

[C88b] Chen, P.: Empirical and theoretical evidence of economic chaos September 7, 1987 System Dynamics Review, vol. 4, nr. 1–2 (1988)

[C93] Chen, P.: Searching for Economic Chaos: A Challenge to Econometric Practice and Nonlinear Tests, Day, R H. and P. Chen., (eds), Nonlinear Dynamics and Evolutionary Economics, 217–253, Oxford University Press. (1993)

[CL93] Cromwell, J.B. and W.C. Labys.: Testing for Nonlinear Dynamics and Chaos in Agricultural Commodity Prices, Memo Institute for Labor Study, West Virginia University (1993)

[D94] Day, R.H.: Complex Economic Dynamics, vol 1. An introduction to Dynamical Systems and Market Mechanisms, Cambridge, MA: MIT Press (1994)

[F05] Faggini, M.: Un Approccio di Teoria del Caos all'Analisi delle Serie Storiche Economiche, Rivista di Politica Economica, luglio agosto (2005)

[F06a] Faggini, M.: Visual Recurrence Analysis: an application to economic time series in Complexity Hints for Economic Policy (eds) M. Salzano e D. Colander, Springer Edition - forthcoming (2006a)

[F06b] Faggini, M.: Chaotic Systems and New Perspectives for Economics Methodology. A Review from Literature forthcoming in Complexity and Policy Analysis: Decision Making in an Interconnected World, Greenwich, CT: IAP Publishing a cura di Dennard, L., K.A. Richardson and G. Morcol (2006b)

[FS88] Frank, M. and T. Stengos: Chaotic Dynamics in Economic Time Series, Journal of Economic Surveys, vol. 2 n° 2, 103–133 (1988)

[FS89] Frank, M. and T. Stengos: Measuring the Strangeness of Gold and Silver Rates of Return, Review of Economic Studies, vol. 56, 553–567 (1989)

[G90] Goodwin, R. M.: Chaotic Economic Dynamics, Clarendon Press, Oxford (1990)

[GT92] Granger, C. W. J. and T. Terasvirta: Experiments in Modeling Nonlinear Relationships Between Time Series, in Martin Casdagli and Stephen Eubank, eds., Nonlinear Modeling and Forecasting, Diagnostic Testing for Nonlinearity, Chaos, and General Dependence in Time Series Data. Redwood City, California: Addison–Wesley, 189–198 (1992)

[H91] Hommes, C. H.: Chaotic Dynamics in Economics Models. Some Single Case Studies, Gronigen, The Netherlands, Wolthers-Nordhoff (1991)

[Hs91] Hsieh, D. A.: Chaos and nonlinear dynamics: application to financial markets, Journal of Finance, 46, issue 5, 1839–1877 (1991)

[K04] Kuchta, S.: Nonlinearity and Chaos in Macroeconomics and Financial Markets, working paper University of Connecticut (2004)

[M92] Medio, A.: Chaotic Dynamics: Theory and Application to Economics. Great Britian (1992)

[P91] Peter E.E.: A chaotic attractor for the S.eP. 500, Financial Analysis Journal, 47, 55–62 (1991)

[P95] Potter, S.: A nonlinear approach to U.S. GNP, Journal of Applied Econometrics, 10, 109–125 (1995)

[Pr01] Prokhorov A. B.: Nonlinear Dynamics and Chaos Theory in Economics: a Historical Persective, December, Work in progress, http://www.msu.edu/user /prohorov/ (2001)

[P97] Puu, T.: Nonlinear Economics Dynamics, Berlin, Springer (1997)

[T92] Tsonis, A. A.: Chaos: From Theory to Applications. Plenum Press. New York and London (1992)

[WL98] Wei, A. and R.M. Leuthold.: Long Agricultural Futures Prices: ARCH, Long Memory, or Chaos processes?, OFOR Paper, n. 03 May (1998)

[YB93] Yang, S. R. and B.W. Brorsen.: Nonlinear Dynamics of Daily Futures Prices: Conditional Heteroschedasticity or Chaos?, The Journal of Futures Markets, vol. 13, 175–191 (1993)

Generalized Influence Functions and Robustness Analysis

Matteo Fini and Davide La Torre

Summary. The notion of influence function was introduced by Hampel and it plays a crucial role for important applications in robustness analysis. It is defined by the derivative of a statistic at an underlying distribution and it describes the effect of an infinitesimal contamination at point x on the estimate we are considering. We propose a new approach which can be used whenever the derivative doesn't exist. We extend the definition of influence function to non-smooth functionals using a notion of generalized derivative. We also prove a generalized von Mises expansion.

Key words: Robustness analysis; Influence function; Gross-error sensitivity; Prohorov distance; Qualitative robustness; Non-smooth analysis.

1 Introduction

The idea of influence function, recently proposed by Hampel in 1974, can be considered a crucial notion for the important role it plays in robustness and stability analysis (Hampel, Ronchetti, Rousseeuw and Stahel, 1986). The robustness analysis using influence functions is strongly based on the existence of the derivative of a functional and this opened the way to the use of different concepts of derivatives ([Ha74]). Essentially, three types of generalized derivative are usually used for robustness analysis: Gateaux, compact and Frechet derivatives. Fernholz (1995) presents an interesting discussion on mathematical treatment of these concepts from a statistical perspective. Gateaux derivative is a very important concept in robustness analysis, it can be computed for many estimators and test statistics and can be used to introduce new robust statistical procedures ([Cl86]). On the other hand, the notion of Frechet differentiability is a stronger concept which implies the existence of Gateaux derivative. It guarantees the asymptotic normality of the corresponding statistic ([Be93] and [Cl86]) and is used to get stability properties under small deviations of the model. In Clarke (1986) some results which imply Frechet differentiability for a large class of estimators are shown. Frechet differentiability is also essential in other fields in statistics (for example the bootstrap).

Cira Perna, Marilena Sibillo (Eds.): Mathematical and Statistical Methods in Insurance and Finance

The aim of this work is to propose a new approach which can be used when Gateaux derivative doesn't exist. In this paper we introduce a generalized definition of influence function and we prove a generalized Von Mises theorem. This new definition of influence function is based on a notion of first order generalized derivative. We consider the definition of generalized Dini directional derivative for scalar and vector functionals; this type of derivative is built by taking the set of all cluster points of sequences of particular incremental ratios. The notion of influence function is strictly connected to the robustness issue. In stability analysis a classical definition is the concept of *qualitative robustness* which is based on the notion of Prohorov distance ([Pr56]). The richest quantitative robustness information is provided by the influence function. It describes the effect of an infinitesimal contamination at point x on the estimate and it can be used to analyze the relative influence of individual observations.

The paper is structured as follows. In the second and third sections we recall the notion of qualitative robustness, the classical definition of influence function and a notion of measure which is used in the definition of *B-robustness*. In the fourth section we will recall some mathematical tools which will be useful for the extensions of the fifth section.

2 Prohorov Distance and Qualitative Robustness

Let X be a separable complete metric space, B a Borel σ-algebra and d a distance. For $\epsilon > 0$ and $A \subset X$ we define $A^{\epsilon} = \{x \in X : d(x, A) < \epsilon\}$ with $d(x, A) = \inf_{z \in A} d(x, z)$. Let F and G be two measures on the measure space (X, B) and let $\pi(F, G) = \inf\{\epsilon > 0 : F(A) < G(A^{\epsilon}) + \epsilon, \forall A \in B\}$. The Prohorov distance between F and G is defined by $\pi(F, G) = \max\{\pi(F, G), \pi(G, F)\}$. Using this notion of distance, Hampel introduced the definition of qualitative robustness.

Definition 1. *Let T_n be a sequence of estimator and F a distribution function. We say that T_n is qualitative robust in F if, for all $\epsilon > 0$, there exists $\delta > 0$ such that, for all $EF \in \mathcal{F}(X)$ and for all n, we have*

$$\pi(F, EF) < \delta \Rightarrow \pi(LD_F(T_n), LD_{EF}(T_n)) < \epsilon,$$

where $LD_F(T_n)$ is the distribution law of T under F and $LD_{EF}(T_n)$ is the distribution law of T under EF.

3 Influence Function and B-robustness

Let T be a statistics of the real parameter $\theta \in \Theta$, where $\Theta \subset \mathbb{R}$ is an open and convex set of real numbers. In other words, T is a functional of the i.i.d. sample $(X_1, X_2, \ldots X_n)$, where $(\Omega, \mathcal{F}, p)$ is a probability space and $X_i : \Omega \to X$ is a set of random variables, $X_i \sim F(\cdot, \theta)$. Instead of relying on this data, we could use the

distribution of the random variables. We will now see what happens to an estimator when we change the distribution of the data slightly. The importance of IF lies here, in its heuristic interpretation.
IF describes the infinitesimal behavior of the asymptotic value and so measures the asymptotic bias caused by contamination in the observations. So, let G now be some distribution in $dom(T)$. The question is, what happens when the data doesn't follow the model F exactly but another, slightly different, say G? What we need is a notion of derivative of T in the direction $G - F$. In order to introduce this notion, consider T as functional on the space of distribution functions and let $dom\ (T)$ be the domain, that is the non-empty convex subset $\mathcal{F}\ (X)$ of the set of all probability measures built on X. We will write $T : \mathcal{F}(X) \to \mathbb{R}$.

Definition 2. *A functional T, defined on a probability measure, is differentiable at F in the direction $G - F$ if there exists the following limit*

$$T'(F; G - F) = \lim_{t \to 0} \frac{T\left((1-t)\,F + tG\right) - T\,(F)}{t}\,. \tag{1}$$

Definition 3. *Let T be a differentiable functional at F in the direction $G - F$. We say that it is compactly differentiable if there exists a real function $r : X \to \mathbb{R}$ such that, for all G (with $G \in dom\ (T)$) we have*

$$T'(F; G - F) = \int r\,(x)\,dG\,(x)\ . \tag{2}$$

Now replacing G with the Dirac's Delta as a probability measure, we get the following definition.

Definition 4. *Given a functional $T : \mathcal{F}(X) \to \mathbb{R}$ the influence function IF at a point $x \in X$ is defined as*

$$IF\,(x; T, F) = \lim_{t \downarrow 0} \frac{T\left((1-t)\,F + t\delta_x\right) - T\,(F)}{t}\,, \tag{3}$$

in such $x \in X$ where the limit exists.

Considering Taylor expansion of T in F, we get

$$\begin{aligned} T\,(G) &= T\,(F) + T'_F\,(G - F) + \mathit{reminder} \\ &= T\,(F) + \int IF\,(x; T, F)\,d\,(G - F)\,(x) + \mathit{reminder}. \end{aligned} \tag{4}$$

Now let $G = F_n$, where F_n is the empirical distribution function. Computing this we get

$$\begin{aligned} T\,(F_n) &= T\,(F) + \int r(x)d\,(F_n - F)\,(x) + \mathit{reminder}\,(F_n - F) \\ &= T\,(F) + \int r(x)dF_n\,(x) + \mathit{reminder}\,(F_n - F)\,, \end{aligned}$$

since $\int r(x)dF(x) = 0$. This expression is called "von Mises expansion" of T in F. The linear term is described by the expression $\int r(x)dF_n(x) = \frac{1}{n}\sum_{i=1}^{n} r(X_i)$ and so the previous Taylor expansion can be rewritten as

$$\sqrt{n}\,(T(F_n) - T(F)) = \frac{1}{\sqrt{n}}\sum_{i=1}^{n} r(X_i) + \sqrt{n}\,reminder\,(F_n - F)\,. \quad (5)$$

The following theorem can be found in [Fe95].

Theorem 1. *Suppose* $0 < Er^2(X) = \sigma^2 < \infty$ *and* $\sqrt{n}\,reminder\,(F_n - F) \xrightarrow{P} 0$. *Then* $\sqrt{n}\,(T(F_n) - T(F)) \xrightarrow{D} N(0, \sigma^2)$ *when* $n \to \infty$.

In literature, this is called von Mises method. What are the properties for a "good" influence function? The influence function describes the effect on T performed by an infinitesimal perturbation of the data at point x. From the robustness point of view, Hampel introduced some important summary values of the IF:

- Rejection point

$$\rho^* = \inf_{r>0}\{r : IF(x; T, F) = 0, |x| > r\}\,. \quad (6)$$

- Gross-error sensitivity

$$\gamma^*(T, F) = \sup_{x} |IF(x; T, F)|\,. \quad (7)$$

- Local-shift sensitivity

$$\lambda^*(T, F) = \sup_{(x,y)\in X\times X, x\neq y} \left| \frac{IF(y, T, F) - IF(x, T, F)}{y - x} \right|\,. \quad (8)$$

The gross-error sensitivity measures the worst influence that a small amount of contamination of fixed size can have on the estimator T. The local-shift sensitivity, which looks a lot like a lipschitz constant, represents the effect of shifting an observation slightly from x to a closed point y. The desirable properties of an influence function are: small gross-error sensitivity, small local-shift sensitivity and finite rejection point. Using this definition of influence function, in literature many definitions of stability have been proposed. The following recalls the notion of B-robustness.

Definition 5. *We say that* T *is B-robust if* $\gamma * (T, F) < \infty$.

Example 1. Let $X \sim N(0, 1)$ be a random variable, θ an unknown parameter, $F(x)$ be a distribution function of X. Let $F_\theta(x) = F(x - \theta)$ and let $\widetilde{\theta}$ be the real value of the parameter to be estimated. Let $\overline{X} = \int xF(x)\,dx$ be the estimator T we are considering. The influence function is given by

$$IF\left(x; \overline{X}, F\right) = \lim_{t\downarrow 0} \frac{\int u\,[(1-t)F(x) + t\delta_x]\,du - \int uF(x)\,du}{t} \quad (9)$$

$$= \lim_{t\downarrow 0} \frac{(1-t)\int uF(x)\,du + t\int u\delta_x du - \widetilde{\theta}}{t} = x - \widetilde{\theta}\,. \quad (10)$$

The value of gross-error sensitivity is

$$\gamma * (T, F) = \sup_{x: IF(x;T,F)} |IF(x; T, F)| = \sup_x \left|x - \widetilde{\theta}\right| = \infty ,$$

so the mean is not a B-robust estimator. The local-shift sensitivity is

$$\lambda * (T, F) = \sup_{x \neq y: \exists IF(x;T,F), IF(y;T,F)} \frac{|IF(y; T, F) - IF(x; T, F)|}{|y - x|} = 1 ,$$

while the rejection point is

$$\rho * (T, F) = \inf\{r > 0 : IF(x; T, F) = 0, \forall |x| > r\} = \infty .$$

4 Generalized Derivatives for Scalar and Vector Functions

All definitions and notions in previous section are based on the notion of differentiability. The existence of such a notion of derivative is not always guaranteed and so it is natural to ask what happens if this limit doesn't exist. We will extend the definition of influence function using the notion of first order Dini generalized derivative; we will use a generalized mean value theorem and Taylor expansion in order to prove a generalized von Mises expansion. We consider the case of function $f : \mathbb{R}^n \to \mathbb{R}^m$ since introducing a notion of generalized directional derivative of T at point x in the direction d, one can regard $\phi(t) = T(x + td)$. A function $f : \mathbb{R}^n \to \mathbb{R}^m$ is said to be locally lipschitz at x_0 if a constant K_f exists such that $\|f(x) - f(y)\| \leq K_f \|x - y\|$, for all $x, y \in U(x_0)$.

Definition 6. *Let $f : \mathbb{R}^n \to \mathbb{R}$ be a given function and $x_0 \in \mathbb{R}^n$. For such a function, the definition of upper Dini generalized derivative $\overline{f}'_D$ at x_0 in the direction $u \in \mathbb{R}^n$ is*

$$f'_D(x_0; u) = \limsup_{t \downarrow 0} \frac{f(x_0 + tu) - f(x_0)}{t} . \tag{11}$$

In an analogous way, one can define a lower derivative. The following result states a generalized mean value theorem for f ([Di81]).

Lemma 1. *Let $f : \mathbb{R}^n \to \mathbb{R}$ be a locally lipschitz function, then $\forall a, b \in \mathbb{R}^n$, $\exists \alpha \in [a, b]$ such that*

$$f(b) - f(a) \leq \overline{f'}_D(\alpha; b - a) . \tag{12}$$

These results can be extended to vector functions.

Definition 7. *Let $f : \mathbb{R}^n \to \mathbb{R}^m$ be a given function and $x_0 \in \mathbb{R}^n$. For such a function, the definition of upper Dini generalized derivative f'_D at x_0 in the direction $u \in \mathbb{R}^n$ is*

$$f'_D(x_0; u) = \left\{ l = \lim_{n \to +\infty} \frac{f(x_0 + t_n u) - f(x_0)}{t_n}, t_n \downarrow 0 \right\} . \tag{13}$$

Obviously this set is non-empty when locally lipschitz functions are considered. In fact, if f is locally lipschitz at x_0 then $f'(x_0; d)$ is a non-empty compact subset of $\mathbb{R}^m$. If $f(x) = (f_1(x), \ldots, f_m(x))$ then from the previous definition it is not difficult to prove that $f'_D(x_0; u) \subset (f_1)'_D(x_0; u) \times \ldots (f_m)'_D(x_0; u)$. We now show that this inclusion can be strict. Let us consider the function $f(x) = (x\sin(x^{-1}), x\cos(x^{-1}))$ for which we have $f'_D(0; 1) \subset \{d \in \mathbb{R}^2 : \|d\| = 1\}$ while $(f'_1)_D(0; 1) = (f'_1)_D(0; 1) = [-1, 1]$. The following result can be found in [La04].

Theorem 2. *Let $f : \mathbb{R}^n \to \mathbb{R}^m$ be a locally Lipschitz vector function. Then the following generalized mean value theorem holds*

$$0 \in f(b) - f(a) - clconv\left\{f'_D(x; b-a) : x \in [a, b]\right\}. \tag{14}$$

5 Generalized Influence Functions and Generalized B-robustness

We now introduce two definitions of generalized influence functions (for scalar and vector functionals) based on the notions of generalized derivatives that we have considered in the previous section.

5.1 Scalar case

The following definition states a notion of generalized influence function when scalar functionals are considered.

Definition 8. *A functional T, defined on a probability measure, is upper compactly differentiable on the measure F if there exists a real function $r^+ : X \to \mathbb{R}$ such that, for all G (with $G \in dom(T)$) we have*

$$\overline{T}'_D(F, G-F) = \limsup_{t\downarrow 0} \frac{T((1-t)F + tG) - T(F)}{t} = \int_X r^+(x)\, dG(x). \tag{15}$$

T is lower compactly differentiable on the measure F if there exists a real function $r^- : X \to \mathbb{R}$ such that, for all G (with $G \in dom(T)$) we have

$$\underline{T}'_D(F, G-F) = \liminf_{t\downarrow 0} \frac{T((1-t)F + tG) - T(F)}{t} = \int_X r^-(x)\, dG(x). \tag{16}$$

Definition 9. *Given a functional $T : \mathcal{F}(X) \to \mathbb{R}$ the generalized upper influence function IF^+ at a point $x \in X$ is defined as*

$$IF^+(x; T, F) = \limsup_{t\downarrow 0} \frac{T((1-t)F + t\delta_x) - T(F)}{t}, \tag{17}$$

while the generalized lower influence function IF^- at a point $x \in X$ is defined as

$$IF_-(x; T, F) = \liminf_{t\downarrow 0} \frac{T((1-t)F + t\delta_x) - T(F)}{t}. \tag{18}$$

Now replacing IF with the Dirac's Delta as probability measure, we obtain $IF^{+}(x; T, F) = r^{+}(x)$ and $IF^{-}(x; T, F) = r^{-}(x)$. It is trivial to prove that $IF_{-}(x; T, F) \leq IF_{+}(x; T, F)$. If $\widetilde{F} = \left(1 - \widetilde{t}\right) G + \widetilde{t} F$ using a generalized mean value theorem, we get $T(G) - T(F) \leq \overline{T}'_D(\widetilde{F}, G - F)$. Now it is possible to prove the following theorem which states a generalized von Mises expansion for IF^{+}.

Theorem 3. *Let $F \to \overline{T}'_D(F, G - F)$ be upper semicontinuous, then $\forall \epsilon > 0$, there exists $\delta > 0$ such that $\forall G \in U(F, \delta)$ and we have*

$$T(G) - T(F) \leq \int_X IF^{+}(x; T, F)\, dG(x) + \epsilon . \tag{19}$$

5.2 Vector case

We can now extend the notion of influence function to vector non-smooth functionals. This notion requires the definition of multifunction. A multifunction $F : X \rightrightarrows Y$ is a function from X to the power set 2^Y. If $F(x)$ is closed, compact or convex, we say that F is a closed, compact or convex value, respectively. A function $f : X \to Y$ is a selection of F if $f(x) \in F(x)$ for all $x \in X$. In the following, we consider the notion of integral of a multifunction in Aumann sense, that is

$$\int F(x) d\mu(x) = \left\{ \int f(x) d\mu(x),\ f(x) \in F(x), \forall x \in X \right\} .$$

Definition 10. *A vector functional T, defined on a probability measure, is compactly differentiable on the measure F if there exists a multifunction $r(x) : X \to 2^{\mathbb{R}^m}$ such that*

$$T'_D(F, G - F) = \lim_{t_n \downarrow 0} \frac{T(1 - t_n)G + t_n F) - T(F)}{t_n} = \int_X r(x) dG(x) . \tag{20}$$

Definition 11. *Given a functional $T : \mathcal{F}(X) \to \mathbb{R}^m$ the generalized influence function GIF at a point $x \in X$ is defined as*

$$GIF(x; T, F) = \left\{ l \in \mathbb{R}^m : l = \lim_{t_n \downarrow 0} \frac{T(1 - t_n)F + t_n \delta_x) - T(F)}{t_n} \right\} . \tag{21}$$

If $G = \delta_x$ then we get $GIF(x) = r(x)$. Using a generalized mean value theorem we get

$$T(G) - T(F) \in \operatorname{cl\,conv} \left\{ T'_D(\widetilde{F}, G - F),\ \widetilde{F} = \left(1 - \widetilde{t}\right) F + \widetilde{t} G, \widetilde{t} \in (0, 1) \right\} . \tag{22}$$

The following result states a generalized von Mises expansion.

Theorem 4. *Let $F \to T'_D(F, G - F)$ be upper semicontinuous. Then for all $\epsilon > 0$ there exists $\delta > 0$ such that $\forall G \in U(F, \delta)$ and we have that*

$$T(G) - T(F) \in \int_X GIF(x; T, F)\, dx + \epsilon B(0, 1) . \tag{23}$$

The previous definitions of generalized influence functions can be used to extend the notion of B-robustness.

Definition 12. *Let $T : \mathcal{F}(X) \to \mathbb{R}$ be a scalar non-smooth functional. We say that T is generalized B-robust if* $\sup_{x \in X} IF^{+}(x; T, F) < +\infty$. *If T is a vector non-smooth functional, T is said to be generalized B-robust if the set $\bigcup_{x \in X} GIF(x; T, F)$ is compact.*

References

[Be93] Bednarski, T.: Frechet Differentiability of Statistical Functionals and Implications to Robust Statistics. In: Morgenthaler, S., Ronchetti, E., Stahel, W. A. (eds.) New Directions in Statistical Data Analysis and Robustness. Birkhauser, Basel, **83**, 23–34 (1993)

[Cl86] Clarke, B. R.: Nonsmooth Analysis and Frechet Differentiability of M Functionals. In: Probability Theory and Related Fields, **73**, 197–209 (1986)

[Di81] Diewert, W. E.: Alternative characterizations of six kinds of quasiconcavity in the nondifferentiable case with applications to nonsmooth programming. In: Schaible S., Ziemba, W. T. (eds.) Generalized Concavity in Optimization and Economics. Springer, 51–95 (1981)

[La04] Cusano, C., M. Fini, M., La Torre, D.: Characterization of convex vector functions and optimization. Journal of Inequalities in Pure and Applied Mathematics, **5**, art. 101 (2004)

[Fe95] Fernholz, L. T.: Von Mises calculus for statistical functionals. Lecture Notes in Statistics, Springer-Verlag, New York (1995)

[Ha74] Hampel, F. R.: The influence curve and its role in robust estimation. Journal of the American Statistical Association, **69**, 383–393 (1995)

[Ha86] Hampel, F. R., Ronchetti, E. M., Rousseeuw, P. J., Stahel, W. A.: Robust Statistics: The Approach Based on Influence Functions. Wiley, New York (1986)

[Hu81] Huber, P. J.: Robust Statistics. Wiley, New York (1981)

[Pr56] Prohorov, Y. V.: Convergence of random processes and limit theorems in probability theory. Theoretical Probability and Applications, **1**, 157–214 (1956)

[Ro81] Rousseeuw, P., Ronchetti, E.: Influence curves of general statistics. Journal of Computational and Applied Mathematics, **7**, 161–166 (1981)

Neural Networks for Bandwidth Selection in Non-Parametric Derivative Estimation

Francesco Giordano and Maria Lucia Parrella

Summary. In this paper we consider the problem of bandwidth selection in local polynomial estimation of derivative functions. We use a dependent data context, and analyze time series which are realizations of strictly stationary processes. We consider the estimation of the first derivative of the conditional mean function for a non-linear autoregressive model. First of all, we emphasize the role assumed by the smoothing parameter, by showing how the choice of the bandwidth is crucial for the consistency of the non-parametric estimation procedure, through an example on simulated data. We then use a new approach for the selection of such a parameter, based on the neural network technique. Such alternative method presents several advantages with respect to the traditional approach used so far.

Key words: Bandwidth selection; Dependent data; Local polynomial estimators; Neural networks.

1 Introduction

Let $\{X_t; t = 1, \ldots, n\}$ be a realization of length n from a real valued stationary stochastic process, $\{X_t; t \in \mathbb{N}\}$. We consider the following autoregressive model

$$X_t = m\left(X_{t-1}\right) + \varepsilon_t, \tag{1}$$

where the errors $\{\varepsilon_t\}$ are *i.i.d.* with mean equal to 0 and variance $\sigma^2 \in (0, +\infty)$. Model (1) is useful to model non-linear structures commonly encountered in financial and econometric time series ([Tjo94]). Also useful in economics and finance is the estimation of derivative functions: for example, option pricing, delta hedging, sensitivity analysis and optimization problems etc. Sometimes the parametric form of such derivative functions is unknown, unless we make some strong assumptions. It is useful, therefore, to have a non-parametric tool to estimate the derivatives.

The conditional mean function $m(\cdot)$ is supposed to have a continuous third order derivative on $\mathbb{R}$. Moreover, $\lim_{|x|\to\infty} \frac{m(x)}{x} = c$, $|c| < 1$. The error terms ε_t have continuous and positive density function.

Cira Perna, Marilena Sibillo (Eds.): Mathematical and Statistical Methods in Insurance and Finance

The models in (1) have a markovian representation. Under the abovementioned conditions, the process $\{X_t\}$ is geometrically ergodic and exponentially α-mixing ([FKMN02, Dou94)]).

The easiest way to estimate non-parametrically the function $m(x)$ and its derivatives is by using Local Polynomial estimators. Such estimators have good theoretically properties, as shown by [FG96]. Nevertheless, they are subject to the correct preliminary choice of a smoothing parameter, the bandwidth of the kernel function, which in fact may compromise considerably their performance. As a result, the fundamental reason for using these non-parametric tools, i.e. robustness against a parametric mis-specification, vanishes completely. In this paper we highlight this problem by showing empirically the difficulties in selecting the correct bandwidth, through an example on simulated data. Then we describe and use a new method for the selection of such parameter which is based on the neural network technique. Such method has been proposed and analyzed in a different paper, and here we refer to the results reported therein ([GP06]). The proposed method presents several advantages with respect to the traditional approach, as will be summarized later on in this paper.

2 Local Polynomials for Non-parametric Derivative Estimation

For simplicity, let us consider the local polynomial estimator of the first derivative function $m'(x)$, for $x \in \mathbb{R}$. We consider the local quadratic estimator, which is usually used for the estimation of the first derivative of a function. There are theoretical reasons for using such particular kind of local polynomial estimator (see, for example, [FG96]). This estimator is given by solving the following minimization problem

$$\arg\min_{\beta_0,\beta_1,\beta_2} \sum_{t=2}^{n} \left(X_t - \sum_{j=0}^{2} \beta_j(x)\,(X_{t-1}-x)^j \right)^2 K\left(\frac{X_{t-1}-x}{h_n}\right),$$

with respect to $\beta_0(x)$, $\beta_1(x)$ and $\beta_2(x)$. Note that $\hat{m}_h(x) \equiv \hat{\beta}_0(x)$, $\hat{m}'_h(x) \equiv \hat{\beta}_1(x)$ and $m''_h(x) \equiv 2\hat{\beta}_2(x)$. The function $K(u)$, called the kernel function, is a density function definite on the compact support $[-1, 1]$. This function is influenced by the smoothing parameter h_n, which is called the *bandwidth* of the kernel function. It is clear that h_n represents a scale parameter for the weigth function $K(\cdot)$, so it determines the amount of local averaging and the smoothness of the estimated function. A bandwidth h_n of order $O(n^{-1/7}) = C_n n^{-1/7}$ is required in order to obtain the consistency and the asymptotic normality of the estimator $m'_h(x)$, where C_n is a positive constant ([MF97]). It is of prime importance that the value of C_n is correctly identified, since even a small variation of the bandwidth may affect the results of the estimation. For example,

- A value of h_n that is too large may cause an estimation which is remarkably biased and therefore unreliable.
- A value of h_n that is too small may cause a substantial increase in the variance of the estimations and therefore the estimator tends to be inefficient.

- Uncorrect values of h_n may cause confidence intervals which are too large and/or completely misleading. This also reflects on the hypothesis tests which are built on such kind of estimators.

Finally, the optimal choice of the smoothing parameter must be made by taking into account the opposite need of controlling bias and variance of the estimations. This is not such a trivial thing to do, so it is useful to have an automatic selection procedure which takes into account this *trade-off*. Much research has been done on data driven bandwidth selection methods.

3 The Selection of the Smoothing Parameter

The methods proposed so far in the literature for the selection of the optimal bandwidth may be divided into two broad categories: *cross-validation* methods and *plug-in* methods. Both of them are based on the application of some optimality criteria, generally the minimization of an appropriate measure of the estimation error. Some references on this topic, for a dependent data context, are [HarVie92, RupAlt95, HalAlt95, Har96, Ops97, KimCox97, MasFan97, Sko00 and FraAlt04].

Each one of these approaches has its own particular strengths and weaknesses. Nevertheless, they have been compared by several authors, which show evidence of a substantial superiority of plug-in procedures over cross-validation ones. Their conclusions are based on real data examples, simulation studies and asymptotic theory (see for example, [Loa99, RSW95, PT92, HSJM91 and Chi91]).

In a parallel work ([GP06 and GP06b]), we propose a new plug-in method for the selection of h_n based on the use of the neural network technique. Here we apply this method to the problem of estimating the derivative function $m'(x)$. Let us consider the following quantity

$$\widetilde{MISE}(h) = \int \left\{Bias^2\left[\hat{m}'_h(x)\right] + \text{Var}\left[\hat{m}'_h(x)\right]\right\} w(x;\tau) f_X(x) dx, \qquad (2)$$

where $\hat{m}'_h(x)$ is the local quadratic estimator of the function $m'(x)$, $f_X(x)$ is the density of X_t, while the weight function $w(x;\tau)$ is taken as the density of a normal distribution with mean equal to zero and variance equal to the variance of the process $\{X_t\}$. (2) represents a modified version of the mean integrated square error, and so represents a global measure of the estimation error. [MF97] gives the asymptotic bias and the variance of the estimator $\hat{m}'_h(x)$. These can be used to find the global asymptotical optimal bandwidth by minimizing (2):

$$h_{AMISE} = \left(\frac{9\sigma^2 V_1}{2B_1^2 R(m''')}\right)^{1/7} n^{-1/7}, \qquad (3)$$

where $R(m''') = \int [m'''(x)]^2 w(x;\tau) f_X(x) dx$. The quantities B_1 and V_1 are known because they depend only on the kernel function (which is known). On the other hand, the quantities σ^2 and $R(m''')$ are unknown and they must be estimated in some way. Given the ergodicity of the process, we propose the following two estimators

$$\hat{R}_{NN}(m''') = \frac{1}{n-1}\sum_{t=2}^{n}\left[g'''\left(X_{t-1};\hat{\eta}\right)\right]^2 w(X_{t-1};\hat{\tau}); \quad \hat{\sigma}^2_{NN} = \frac{1}{n-1}\sum_{t=2}^{n}\hat{\varepsilon}^2_t. \quad (4)$$

Here $\hat{\tau} = n^{-1}\sum(X_t-\mu)^2$, $\hat{\varepsilon}_t = X_t - g\left(X_{t-1};\hat{\eta}\right)$, $g\left(x;\hat{\eta}\right)$ and $g'''\left(x;\hat{\eta}\right)$ represent respectively the neural network estimation of $m(x)$ and of its third derivative $m'''(x)$. Let us indicate with $\hat{h}_{NN}$ the proposed estimator of the bandwidth, given by plugging into (3) the estimation obtained in (4).

The peculiarity of our approach lies in the fact that it does not require the selection of a preliminary pilot bandwidth, contrary to with the traditional plug-in methods. Plug-in methods are generally implemented as a two stage procedure, where the first stage concerns the selection of a pilot bandwidth $\tilde{h}_n$, which is then used in the second stage to obtain a local polynomial fit of the unknown functionals in (3). In the first stage, the estimation of higher order derivatives is required for the selection of the optimal pilot bandwidth. The main drawback of classic plug-in methods is that they depend heavily on arbitrary specification of such pilot bandwidths, and they fail when this specification is wrong. On the other hand, our method is a one stage procedure, because it does not require the estimation of higher order derivatives. However, it is necessary to identify the number of nodes of the hidden layer, for the estimation of the neural network function. This is an integer parameter which plays the same role as the pilot bandwidth, but its selection is simpler, since it is an integer value of order of some units and not a real value such as the pilot bandwidth. Moreover, there have been proposed in the literature several automatic procedures for the optimal selection of such parameter.

The proposed bandwidth selection procedure may be improved in several directions, and currently this work is in progress. For example, this method gives an estimate of the global optimal bandwidth, that is a bandwidth which is constant over the support of the function. Global bandwidths, however, can lead to undersmoothing or oversmoothing, since they are not sensitive to local curvature changes. Finally, also as an immediate solution to this problem, our method may be implemented as a preliminary bandwidth selector, in a pilot stage of a more refined procedure.

4 An Experiment on Simulated Data

In this section we draw an experiment on simulated data, in order to show empirically the influence that the bandwidth has on the result of the local polynomial estimation, and the difficulties that statistics practitioners have to face to select the correct value of such parameter. Let us consider the following two autoregressive models of order one

$$X_t = 0.8X_{t-1} + \varepsilon_t \quad (5)$$

$$X_t = (0.6 + 0.6e^{-X^2_{t-1}})X_{t-1} + \varepsilon_t. \quad (6)$$

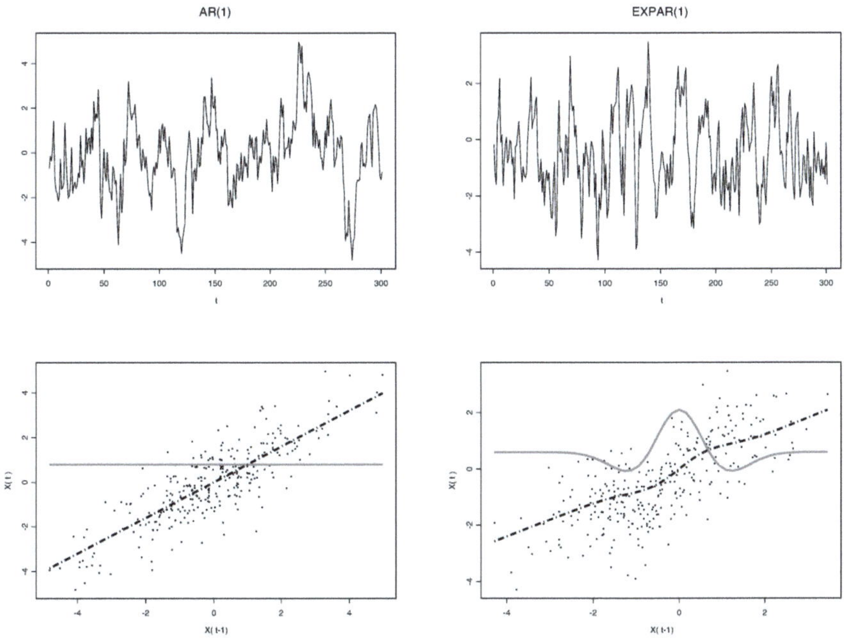

Fig. 1. Top row: time plots of two realizations of length 300 from models (5) and (6). Bottom row: the scatter plots of (X_t, X_{t-1}) for the two time series, together with the conditional mean function (dashed line) and the first derivative of the conditional mean function (solid line).

The first is a linear auroregressive model, while the second is an exponential autoregressive model. It is evident that the conditional mean function of model (5) is equal to $m(x) = 0.8x$, and so it is linear, while the conditional mean function of model (6) is equal to $m(x) = (0.6 + 1.5e^{-x^2})x$, and so it is non-linear.

Fig. 1 shows the time plots and the scatter plots of two realizations of length 300, respectively from model (5) and model (6). As we can see, the two realizations are similar concerning the range of the values and the length of the series. For each realization, we also report the conditional mean function (dashed line) and its first derivative function (solid line). The non-linearity of model (6) appears evident from the behaviour of the first derivative function, more than from the behaviour of the conditional mean function itself.

In practice, one of the main problems faced by the statistics practitioner is the selection of the correct bandwidth. In many cases, it is more difficult to implement an automatic procedure for the selection of such parameter of the estimator than to implement the estimator itself. As a consequence, the selection of the bandwidth is often done by rule of thumb, basing on the practical experience of the analyst, and taking into account the range and the length of the series. This may be misleading, as we will see next.

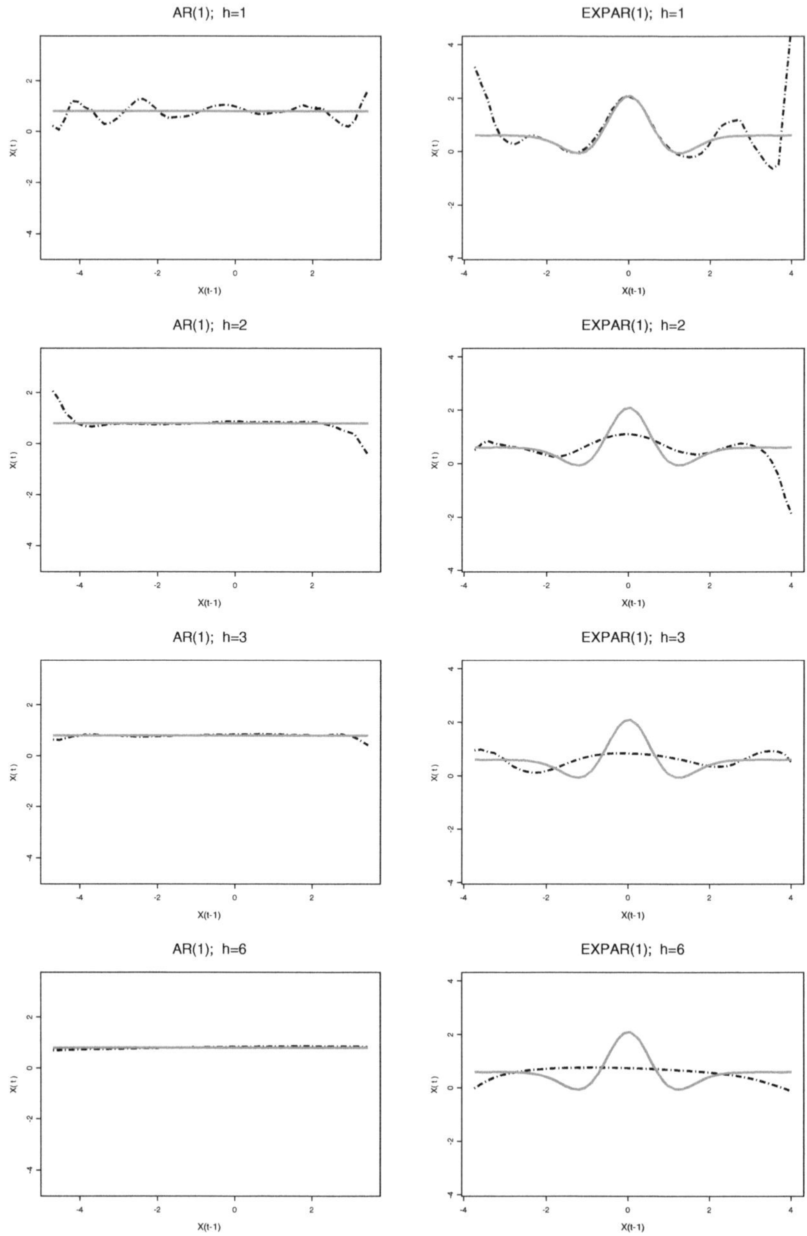

Fig. 2. From top to bottom: Local quadratic estimation of the first derivative of the conditional mean function, for increasing values of the bandwidth h (dashed line). The plots on the left are relative to model (5), while the plots on the right are relative to model (6). In each plot, the true derivative function is represented (solid line).

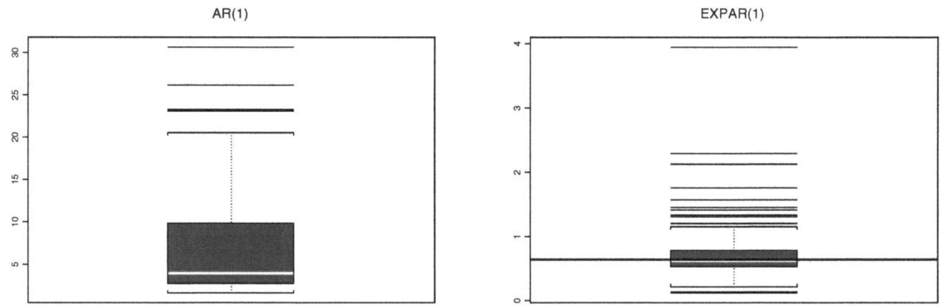

Fig. 3. Box-plots of the estimations of the optimal bandwidth $\hat{h}_{NN}$ obtained with our neural network based procedure, for the 200 replications of length 300 of models (5) and (6). For model (6), the horizontal line indicates the true value of the asymptotical optimal bandwidth h_{AMISE}.

Fig. 2 reports the results of an experiment drawn on the two simulated series. The plots on the left hand side concern the linear model AR, while the plots on the right hand side are relative to the non-linear model EXPAR. In each plot, the dashed line is the local quadratic estimation of the first derivative of the conditional mean function. From top to bottom, we consider increasing values of the bandwidth $h_n = (1, 2, 3, 6)$. The solid line represents the real derivative function. It is evident that the value of h_n has a crucial influence on the results of the estimation. For model AR, little values of h_n give an estimated function which appears to be irregular and. rather distant from the true function (constant), whereas big values of the bandwidth give a good estimation of such first derivative. It is quite the opposite for model EXPAR, for which big values of h_n give an estimated function which appears (erroneously) constant, whereas little values of the bandwidth cause an estimated function which seems rough but nearer to the true function (apart from some boundary effects). This shows that the value of the optimal bandwidth does not depend only on the length and variability of the time series but also (obviously) on some other unknown characteristics, such as the higher order derivatives of the unknown function. It follows that the correct choice of bandwidth is quite difficult, but it is fundamental in order to obtain reliable results from the non-parametric estimation.

The following table reports the results from a simulation study that was carried out in order to test empirically the performance of the neural network bandwidth selection procedure. The first column of the table reports the true values for the optimal global bandwidth, denoted with h_{AMISE}, which has been derived by Monte Carlo simulations. In the other columns of the table there are the quartiles of the distribution of the estimated values of the optimal bandwidth, obtained by considering 200 replications of each model.

	h_{AMISE}	Quartiles of $\hat{h}_{NN}$		
AR(1)	∞	$Q_1 = 2.708$	$Q_2 = 3.895$	$Q_3 = 9.827$
EXPAR(1)	0.642	$Q_1 = 0.535$	$Q_2 = 0.627$	$Q_3 = 0.787$

As we can see, the neural network bandwidth selector $\hat{h}_{NN}$ generates estimates that are satisfactorily near to the value h_{AMISE}, for both models (5) and (6). The median of the bandwidth estimates is close to the true value of the parameter. For model (5), it is easy to show that this value is infinity. As shown in the table, this situation reflects the estimated values of such optimal bandwidth. The results given in the table are represented graphically by the box-plots in Fig. 3.

References

[Chi91] Chiu S. T.: Bandwidth selection for kernel density estimation, *Annals of Statistics*, **19**, 1883–1905 (1991)

[Dou94)] Doukhan P.: *Mixing: properties and examples*, Lecture Notes in Statistics 85. Springer, New York (1994)

[FG96] Fan J., Gijbels I.: *Local polynomial modelling and its applications*, Chapman and Hall, London (1996)

[FOV04] Francisco-Fernández M., Opsomer J., Vilar-Fernández J. M. Plug-in bandwidth selector for local polynomial regression estimator with correlated errors, *Nonparametric Statistics*, **16**, 127–151 (2004)

[FKMN02] Franke J., Kreiss J. P., Mammen E., Neumann M. H.: Properties of the nonparametric autoregressive bootstrap, *Jounal of Time Series Analysis*, **23**, 555–585 (2002)

[GP06] Giordano F., Parrella M. L.: Local polynomial vs neural networks: some empirical evidences, *Proceedings of the National Congress on Time Series Research*, April 18–19, Rome (2006).

[GP06b] Giordano F., Parrella M. L.: Neural Networks for bandwidth selection in local linear regression of time series, *submitted to Computational Statistics and Data Analysis* (2006)

[HLT95] Hall P., Lahiri S. L., Truong Y. K.: On bandwidth choice for density estimation with dependent data, *Annals of Statistics*, **23**, 6, 2241–2263 (1995)

[HSJM91] Hall P., Sheather S. J., Jones M. C., Marron J. S.: On optimal data-based bandwidth selection in kernel density estimation, *Biometrika*, **78**, 263–270 (1991)

[HV92] Härdle W., Vieu P.: Kernel regression smoothing of time series, *Journal of Time Series Analysis*, **13**, 209–232 (1992)

[Har96] Hart J.: Some automated methods of smoothing time-dependent data, *Journal of Nonparametric Statistics*, **6**, 115–142 (1996)

[Loa99] Loader C. R.: Bandwidth selection: classical or plug-in?, *Annals of Statistics*, **27**, 2, 415–438 (1999)

[KC97] Kim T. Y., Cox D. D.: Bandwidth selection in kernel smoothing of time series, *Journal of Time Series Analysis*, **17**, 49–63 (1997)

[MF97] Masry E., Fan J.: Local polynomial estimation of regression functions for mixing processes, *Scandinavian Journal of Statistics*, **24**, 165–179 (1997)

[Ops97] Opsomer J. D.: Nonparametric regression in the presence of correlated errors, in *Modelling Longitudinal and Spatially Correlated Data: Methods, applications and Future Directions*, Springer-Verlag, 339–348 (1997)

[PT92] Park B. U., Turlach B. A.: Practical performance of several data driven bandwidth selectors, *Computational Statistics*, **7**, 251–270 (1992)

[RSW95] Ruppert D., Sheather S. J., Wand M. P.: An effective bandwidth selector for local least squares regression, *Journal of the American Statistical Association*, **90-432**, 1257–1270 (1995)

[Skö00] Sköld M.: A bias correction for cross-validation bandwidth selection when a kernel estimate id based on dependent data, *Journal of Time Series Analysis*, **22**, 4, 493–503 (2000)

[Tjo94] Tjostheim D.: Non-linear time series: a selective review, *Scandinavian Journal of Statistics*, **21**, 4, 97–130 (1994)

Comparing Mortality Trends via Lee-Carter Method in the Framework of Multidimensional Data Analysis

Giuseppe Giordano, Maria Russolillo and Steven Haberman

Summary. In the framework of demographic processes, different approaches can be identified. According to their aims, these approaches can be differentiated in extrapolative and structural methods. The first focus on the homogeneity of trends in order to obtain projection. The second are based on structural models relating demographic variables to other kinds of variables (geographical, social, economical, etc.). Nowadays, this distinction is not so clear and the joint use of explorative and explanatory approaches is increasing. In this paper, we focus on the extrapolative features of the Lee-Carter model (1992) and propose a reading of such method in the framework of the Multidimensional Data Analysis. Our aim is to propose a data analysis strategy exploiting the analytical and geometrical properties of the Lee-Carter method.

Key words: Biplot; Lee-Carter methodology; Mortality forecasting; PCA; SVD.

1 Introduction and Basic Notations

One of the most important techniques to model and forecast mortality is the Lee-Carter model [LC92]. It represents an extrapolative method based on a multiplicative two-factor model used with time series analysis in forecasting.

The model fitting can be addressed from both an algebraic and a statistical point of view. The common tool is the Singular Value Decomposition (SVD). The model used for mortality data is:

$$ln\left(m_{x,t}\right) = \alpha_x + \beta_x k_t + \epsilon_{x,t} \ , \tag{1}$$

which denotes the log of a time series of age-specific death rates $m_{x,t}$ as the sum of an age-specific component α_x, that is independent of time and another component that is the product of a time-varying parameter k_t, reflecting the general level of mortality, and an age-specific component β_x , that represents how, rapidly or slowly, mortality at each age varies when the general level of mortality changes. The $\epsilon_{x,t}$ denotes the error term which undertakes the common hypothesis of the Normal distribution.

Cira Perna, Marilena Sibillo (Eds.): Mathematical and Statistical Methods in Insurance and Finance

The Lee-Carter (LC) method normalizes the estimated parameters by specifying $\sum_t k_t = 0$ and $\sum_x \beta_x = 1$. In other specifications, the method can be identified by different constraints, i.e.: $\sum_x \beta_x^2 = 1$ [GK06].

The demographic literature gives emphasis mainly to the analytical role of the SVD as an estimation method. So far, little attention has been given to the descriptive features of such decomposition. In this paper we exploit both the analytical and descriptive features of the SVD in the light of factorial techniques used as explorative data analysis. In particular, we refer to the Principal Component Analysis (PCA) [Jo02] as a factor reduction method with its particular graphical representation and geometrical interpretation. This method allows a major insight into the demographic data structure and provides several theoretical tools to interpret the underpinning model.

The present paper is organized as follows. The Lee-Carter model in the framework of multidimensional data analysis is introduced in Section 2. The geometric and graphical interpretation of the LC Model is presented in Section 2.1. The last section contains an application to compare Italian mortality rates, divided by age-group and by gender, in the period 1950–2000.

2 The Lee-Carter Model in the Framework of Multidimensional Data Analysis

We can state the LC model (1) referring to the mean centered log-mortality rates:

$$\tilde{m}_{x,t} = ln\left(m_{x,t}\right) - \alpha_x = \beta_x k_t + \epsilon_{x,t}; \quad x = 1, \ldots, X; \; t = 1, \ldots, T. \tag{2}$$

The mean centered log-mortality rates, $\tilde{m}_{x,t}$, can be arranged in the $(T \times X)$ matrix $\tilde{\mathbf{M}}$,

$$\tilde{\mathbf{M}} = \begin{array}{c} \\ 1 \\ 2 \\ \vdots \\ T \end{array} \begin{array}{c} \begin{array}{cccc} 1 & 2 & \ldots & X \end{array} \\ \begin{pmatrix} \tilde{m}_{1,1} & \tilde{m}_{1,2} & \ldots & \tilde{m}_{1,X} \\ \tilde{m}_{2,1} & \tilde{m}_{2,2} & \ldots & \tilde{m}_{2,X} \\ \vdots & \vdots & \tilde{m}_{t,x} & \vdots \\ \tilde{m}_{T,1} & \tilde{m}_{T,2} & \ldots & \tilde{m}_{T,X} \end{pmatrix} \end{array}. \tag{3}$$

The singular value decomposition of the matrix $\tilde{\mathbf{M}}$ can be written as the product of three matrices which have geometric and statistical interpretation. In particular, the SVD model is stated as follows:

$$\tilde{\mathbf{M}}_{T\times X} = \mathbf{S}_{T\times h}\mathbf{V}_{h\times h}\mathbf{D'}_{h\times X} \quad h \leq min\left\{T, X\right\}, \tag{4}$$

where h is the rank of $\tilde{\mathbf{M}}$ and $\mathbf{V}$ is a diagonal matrix of the positive singular values of $\tilde{\mathbf{M}}$. The matrices $\mathbf{S}$ and $\mathbf{D'}$ hold the left and right singular vectors forming an orthogonal basis, respectively, for the rows and the columns of the matrix $\tilde{\mathbf{M}}$. From a statistical point of view, the SVD leads to the Principal Component Analysis. This factorial technique has several properties which give a statistical meaning to the LC model components.

Given the matrix $\tilde{\mathbf{M}}$, the PCA aims at defining new variables (factors) as linear combination of the original columns of such a matrix. This method consists of solving the following characteristic equation:

$$\tilde{\mathbf{M}}'\tilde{\mathbf{M}}\mathbf{u}_j = \lambda_j \mathbf{u}_j \quad j = 1, \dots, h, \tag{5}$$

where u_j is the jth eigenvector associated to the jth eigenvalue λ_j. We stress that λ_j is equal to the squared singular value υ_j and the eigenvectors u_j correspond to the right singular vectors d' forming the rows of the matrix $\mathbf{D}$ (see (4)). Henceforth, the SVD model can be written as:

$$\tilde{\mathbf{M}} = \sum_{j=1}^{p} \upsilon_j s_j d'_j + \sum_{j=p+1}^{h} \upsilon_j s_j d'_j; \quad 1 \leq p \leq h; \quad h \leq min\{T, X\}, \tag{6}$$

where the second summation on the right hand side of equation (6) is the error term which reflects the residual information not captured by the first p components of the SVD approximation.

In equation (7) the correspondence of the models (2) and (6) arises by defining $\beta_x^{<j>} = \upsilon_j d_j$; $k_t^{<j>} = s_j$, where $^{<j>}$ refers to the jth element in the SVD. We also highlight the role of the first component in the decomposition and denote with $\mathbf{E}$ the matrix of the residual terms.

$$\tilde{\mathbf{M}} = \upsilon_1 s_1 d'_1 + \sum_{j=2}^{h} \upsilon_j s_j d'_j = k_t^{<1>} \beta_x^{'<1>} + \mathbf{E}. \tag{7}$$

Empirically, in the estimation phase of the LC model, it is adequate to choose just the first component [Le00]. Nevertheless, some authors have extended the LC model to go beyond the first component [RH03]. From an exploratory point of view, it is possible to verify if a relevant structure in the residual component $\mathbf{E}$ exists. In PCA the following index of Explained Inertia (EI) is usually considered:

$$EI = \frac{\sum_{j=1}^{p \leq h} \lambda_j}{\sum_{j=1}^{h} \lambda_j} \cdot 100, \tag{8}$$

where h is the rank of $\tilde{\mathbf{M}}$, and p is set to 1 in the LC model. If the first p eigenvalues account for the most part of the total variability, the residual term is trivial; hence, the index EI provides empirical and theoretical support for the choice of the value of p to be considered in the model fitting.

2.1 Geometrical and graphical interpretation of the Lee-Carter model

The SVD approximation allows a graphical representation in a reduced subspace of both rows and columns of the matrix $\tilde{\mathbf{M}}$. The geometric reading of such representation is carried out according to the Biplot graphical display [Ga71]. The biplot is a low-dimensional display of a rectangular data matrix. The columns and rows are represented respectively by the column-point coordinates (9) representing the age-groups and by the row-point coordinates (10) representing the years:

$$\varphi_j = v_j d_j = \sqrt{\lambda} u_j = \beta_x^{<j>} \tag{9}$$

$$\psi_j = \tilde{\mathbf{M}} d_j = \tilde{\mathbf{M}} u_j = k_t^{<j>}. \tag{10}$$

The age-group coordinates are shown by drawing a vector from the origin. The coordinates of each point are given by the correlation between the corresponding age-groups and the principal components. The first principal component can be seen as a size variable. It is mainly correlated with those age-groups which give the main contribution to the global inertia since they differentiate from the average trend. Age-groups with similar trends will be represented by vector-points leading to the same direction. The opportunity to represent both row and column coordinates in the same space allows a joint reading of the relationship between age-groups and years. In almost the same way, the position of the years is baricentric with respect to the origin. The year-points close to the origin represent the "average years": moving towards the right (left) side we can find year-points experimenting a mortality greater (lower) than the average.

The main advantage of our approach is the opportunity to show graphically the relationship within the estimated *time varying* parameters k, within the *age-group* coefficients β and between these two sets. Furthermore, we are able to describe the meaningful components of the SVD and explore the information lost in the residual term. Indeed, looking at the components beyond the first one, we could check the pattern of the residuals in order to diagnose some discrepancies from the basic hypotheses.

3 An Application to Italian Mortality Rates in the Period 1950–2000

In order to give an in depth illustration of our procedure, we carry out an application by using death rates for the Italian population, supplied by the *Human Mortality Database*[1] and divided by gender. Death rates consist of death counts divided by the exposure-to-risk. These are estimates of the population exposed to the risk of death measured in terms of person-years.

We consider two data matrices, for males and females, holding the mean centered log-mortality rates. The rows represent *51 years* and the columns *21 age-groups*:$\{0, 1-4, 5-9, \ldots, 95-99\}$. The data can be interpreted as follows: a positive value for each year and each age group indicates a mortality greater than the average value of the correspondent age group. As we can see from Figs. 1 and 2, there is a decreasing trend in each age group, both for males and females. Moreover,

[1] Data available on the web site: *www.mortality.org*

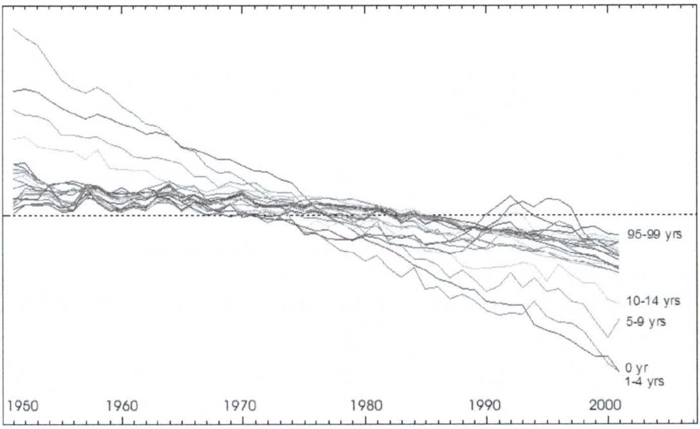

Fig. 1. The mean centered log-mortality rates: Males.

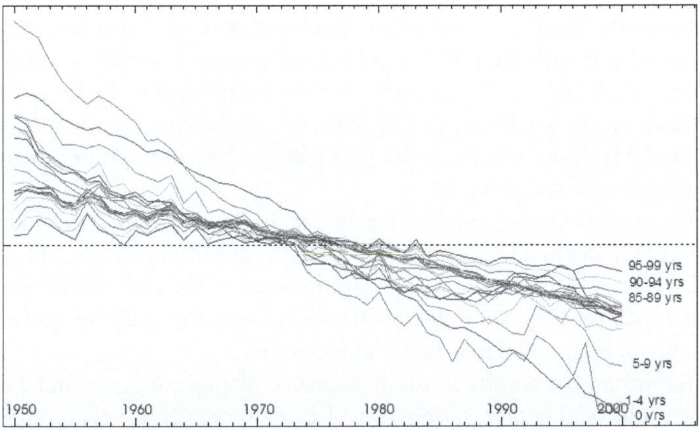

Fig. 2. The mean centered log-mortality rates: Females.

we can observe that the year 1973 represents the transition from positive to negative rates for almost all age groups.

The PCA of both our datasets allows us to estimate and visualize the mortality indices and the age specific components. Firstly, we consider the analysis for males.

As we can see from Table 1, the first eigenvalue shares the larger amount of inertia (95.26%), showing that the residual part is trivial. However, we also consider the second component for descriptive purposes. The second component accounts for 3.28%. The factorial plan explains 98.54% of the total inertia (see Table 1). The same analysis is carried out for females (see Table 2).

In this group we can observe a stronger structure compared to the males. From Table 2, we can see the high percentage (96.98%) of the first component and the very marginal role of the successive ones. As in the previous analysis, we consider the second component with an explorative aim. The factor loadings on the first axis are

Table 1. The first three eigenvalues: Males.

Number	*Eigenvalues*	*Percentage*	*Cumulated Percentage*
1	3.24	95.26	95.26
2	0.11	3.28	98.54
3	0.02	0.49	99.03

Table 2. The first three eigenvalues: Females.

Number	*Eigenvalues*	*Percentage*	*Cumulated Percentage*
1	3.75	96.98	96.98
2	0.08	2.07	99.05
3	0.01	0.33	99.37

higher for the age groups from 0 to 14, decreasing as the age increases. On the second axis, we notice the small value recorder for all age groups. However, age group zero seems to be related with the positive side of the second component, meanwhile age groups 1–4, 25–29 and 30–34 are related to the negative one. As far as the variable-factor correlation is concerned, the leading role of the first component, once more, is highlighted. From the results of the two analyses we can explore the relationship between age groups (see Table 3).

For each age-group we consider the factor loadings and the correlation with the factorial axes. On the first factorial axis, the larger contribution is given by age groups {0, 1–4, 5–9, ..., 95–99}. To stress the significance of the first component in the LC model, we notice the very low values of the factor loading on the second axis. Similar results are given by the variable-factor correlations.

By interpreting the results in the framework of multidimensional data analysis, besides the traditional interpretation of the LC components, we can graphically show the relationship between the estimated time varying parameter k's, within the age-group coefficient β's and between these two sets. The comparison is enhanced by the graphical display of the first principal plan obtained for male and female datasets.

A thorough description of the previously mentioned relationship can be seen in the biplot representation in Figs. 3 and 4. In such a map, each year is represented as a

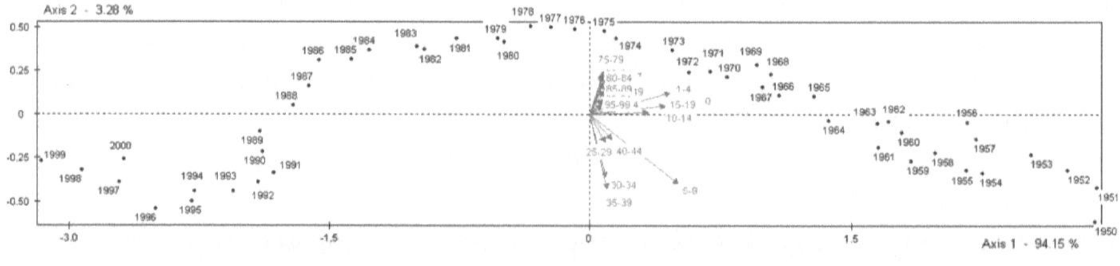

Fig. 3. The first components of the PCA decomposition: Males.

Table 3. Factor loadings of β for the first two components: Males and females.

	Males		*Females*	
Age Group	*Axis 1*	*Axis 2*	*Axis 1*	*Axis 2*
0	0.82	0.04	0.81	0.10
1-4	0.82	0.04	0.93	0.10
5-9	0.90	-0.14	0.61	-0.02
10-14	0.61	0.00	0.48	-0.01
15-19	0.45	0.01	0.37	-0.04
20-24	0.21	0.02	0.42	-0.06
25-29	0.16	-0.05	0.42	-0.10
30-34	0.17	-0.12	0.39	-0.09
35-39	0.18	-0.14	0.37	-0.03
40-44	0.22	-0.05	0.33	0.02
45-49	0.25	0.04	0.29	0.04
50-54	0.25	0.08	0.28	0.05
55-59	0.23	0.09	0.27	0.05
60-64	0.20	0.09	0.29	0.06
65-69	0.16	0.08	0.30	0.05
70-74	0.14	0.08	0.31	0.06
75-79	0.13	0.07	0.29	0.07
80-84	0.14	0.06	0.25	0.07
85-89	0.14	0.04	0.20	0.06
90-94	0.14	0.03	0.14	0.05
95-99	0.12	0.02	0.11	0.04

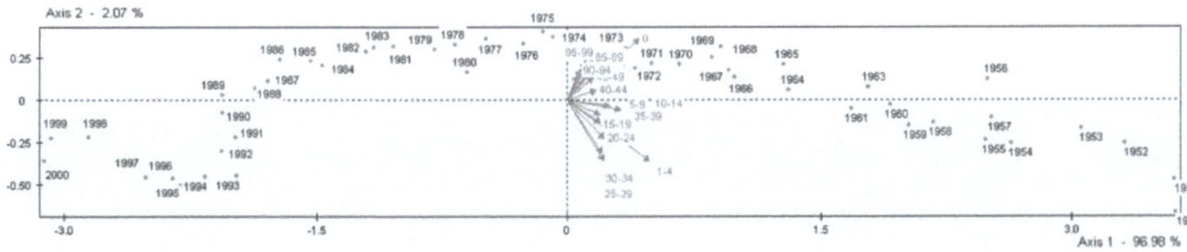

Fig. 4. The first components of the PCA decomposition: Females.

point, whereas the age groups are represented as vectors. The factorial map, provided by the first two components, shows how the first component is basically a factor of trend. Indeed, the year-points lay out from left to right, ranging from the year 2000 to the year 1950, in almost a linear way. Furthermore, the second component has a marginal role in this dynamic. It characterizes a diversification of the mortality indices due to the effect of specific age groups. In particular, on the upper side of the second axis, the eldest age groups are located. On the opposite side of the second factorial axis, we find the central age groups 30–34, 35–39. For males, the choice of just the first component seems to be well supported from the actual data.

References

[Ga71] Gabriel, K. R.: The biplot-graphic display of matrices with application to principal component analysis. Biometrika. **58**, 453–467 (1971)

[GK06] Girosi, F., and King, G.: Demographic Forecasting. Cambridge: Cambridge University Press (2006)

[Jo02] Jolliffe, I. T.: Principal Component Analysis, New York: Springer-Verlag (2002)

[Le00] Lee, R. D.: The Lee-Carter method for forecasting mortality, with various extensions and applications. North American Actuarial Journal, **4**, 1, 80–91 (2000)

[LC92] Lee, R. D., and Carter, L. R.: Modelling and Forecasting U.S. Mortality. Journal of the American Statistical Association, **87**, 659–671 (1992)

[RH03] Renshaw, A. E., and Haberman, S.: Lee-Carter Mortality Forecasting With Age-Specific Enhancement. Insurance: Mathematics and Economics **33**, 255–272 (2003)

Decision Making in Financial Markets Through Multivariate Ordering Procedure*

Luca Grilli and Massimo Alfonso Russo

Summary. One of the main problems in managing multidimensional data for decision making is that it is impossible to define a complete ordering on multidimensional Euclidean spaces. In order to solve this problem, the scientific community has devolped more and more sofisticated tecniques belonging to the wide framework of Multivariate Statistics. Recently some authors [DR04] have proposed an ordering procedure in which the "meaningful direction" is the "worst-best". The aim of this paper is to extend this approach considering that, especially in financial applications, variables are quantified using different scales and, as we will show, this can lead to undesired results. As a matter of fact, we show that, without an appropriate rescaling, variables with a large range of variation (*rv*) are "overweighted" with respect to variables with a small one.

Key words: Multivariate data; Ordering procedures; Normalization; Financial markets.

1 Introduction

A decision maker is usually asked to make decisions faced with many alternatives on the basis of multivariate data. An agent in a financial market faces this problem in decision making and in fact has to determine the portfolio of assets and for each asset has a certain number of different variables to take into due account. Furthermore, the agent has the need to assign each variable an (equal or different) importance according to the tasks. The problem is that, in general, the reality is multidimensional, and it is not always possible to quantify phenomena through quantitative unidimensional "indicator".

The main problem in managing multidimensional data for decision making is that it is not possible to define a complete ordering on multidimensional Euclidean

* Research financially supported by Local Research Project of University of Foggia and National Research Project PRIN, co-financed by MIUR, Italian Ministery for Research, 2004-2006 Titled "Non-linear models in economics and finance: interactions, complexity and forecasting".

Cira Perna, Marilena Sibillo (Eds.): Mathematical and Statistical Methods in Insurance and Finance

spaces. As a result, the scientific community has devolped more and more sofisticated techniques [CRZ99 and Sci98], besides the usual techniques of Multivariate Statistics, in order to solve this important problem in many applications.

The idea underlying these proposals, is that geometry can help multivariate data analysis; in fact data can be considered as a cloud of points in a multidimensional Euclidean space with the main question being: is the interpretation of data on the clouds of points possible?

In principle it seems difficult to answer since clouds of points come from data tables and are not geometric objects; the construction is made on the basis of mathematical structures applied to data. The literature on this subject presents very interesting results such as: Corrispondence Analysis, Principal Component Analysis, Structured Data Analysis and Inductive Data Analysis (complete reference on the subject in [LR04]).

In this paper we consider a different approach to the problem of decision making with multivariate data which answers the following question: Is it possible to reduce multidimensionality to unidimensionality? If this is the case, the problem of ordering multivariate data is completly solved.

One way to answer this question is to find a "meaningful direction" in the data cloud and to consider the projections of data in order to have unidimensional quantities.

In [DR04] the authors provide an ordering procedure, original with respect to classical literature [Bar76], in which the "meaningful direction" is the "worst-best". This approach is very interesting since the "worst-best" direction is meaningful in the sense of contained information.

The aim of this paper is to extend this approach considering a feature that is very important in order to have results that are not dependent on the data set. The question is that, especially in financial applications, data are obtained considering variables in different scales and, as we will show, this can lead to results in which the *rv* of data affects the ordering procedure. As a matter of fact, we demonstrate that, without an appropriate rescaling[2], data with a large *rv* are "overweighted" with respect to data with a small one.

In particular, in financial applications, assets are represented by a certain number of quantitative indicators that are often measured by different scales. For instance, let us suppose that an investor, in order to select assets, takes into account only two indicators: the market capitalization – CAP (in billions of euros) and EPS (in euros). It is quite evident that a variation of a few units in the first variable does not have the same importance as the same variation in the second variable, and this will affect the ordering procedure as we will show with an application in the last section.

[2] In the usual procedure of many statistical methods, his problem is solved through standardization. This makes the variables homogeneous with respect to the mean and variance but not in terms of *rv*.

2 The Ordering Procedure

In ordering procedures in which projections are computed on the "worst-best" direction, the way in which the choice of the worst and best performances is made, say the vectors x_{min} and x_{max}, is very important. This can be done in two different ways: the first is to find the extreme performance *ex-ante* (on the basis of an expert knowledge) and the second is to construct the extreme values through real data. If possible it is better to use the first method to avoid problems rising from the presence of outliers. In the framework of this paper, however, this choice is not crucial.

The main issue is to determine the direction (a vector with norm equal to one) u on which the different values $x_i \in \mathbb{R}^k$ and $i = \{1, \dots, n\}$, are projected and so reduced to a single value. In this case:

$$u = \frac{x_{max} - x_{min}}{\| x_{max} - x_{min} \|}, \tag{1}$$

that is the "worst-best" direction. In order to consider the point x_{min} as the origin of the axis, each vector x_i is transformed into $\tilde{x}_i$ in the following way:

$$\tilde{x}_i = x_i - x_{min}. \tag{2}$$

The projection of the vector $\tilde{x}_i$ on the vector u is the vector:

$$x_i^* = u \cdot \frac{< \tilde{x}_i, u >}{\| u \|} = u \cdot < \tilde{x}_i, u >, \tag{3}$$

where $< \cdot, \cdot >$ is the scalar product. The real number $< \tilde{x}_i, u >$ is the synthetic univariate indicator used in the ordering procedure.

It is important to highlight that the k variables are considered with the same weight and so each one should have the same importance in the performance of the final score, that is considering a vector of weights $p \in \mathbb{R}^k$ with equal components. The decision maker, depending on her tasks, can also consider the case in which the vector of weights has different components according to the importance to be assigned to each variable.

This approach, however, does not consider an important feature that is often present in real data: the problem of data given in different scales. As a result, the parameter with a large rv is overweighted in the final ranking, with respect to the parameter with a low rv [RS06]. A very simple way to confirm this feature is to compute the correlation between the single variables and the final score derived from them. In this case, the correlation is proportional to the range of variation of each variable [RS05] and as a consequence, a vector of weights p^* is used that is different from the desired one that is p.

The previous undesired effect of different ranges of variation, even with homogeneous variability (for instance as a result of standardization), is absolutely not statistically justifiable.

In the next section we propose a possible solution to this problem.

3 The Problem of the Range of Variation

On the basis of the previous analysis it is evident that, without an appropriate transformation of variables, the interpretation of results is strongly influenced. In order to solve this problem, we propose a normalizing transformation of the data before applying the ordering procedure defined by (1) (2) and (3).

In our case it is enough to use [Del95][3]:

$$\hat{x}_i = \frac{x_i - x_i^{min}}{x_i^{max} - x_i^{min}}. \tag{4}$$

This normalization has two important consequences: the *rv* of parameters becomes the same for all data [0, 1] and the direction u has a fixed slope (in the two dimensional case, it is exactly the diagonal of the $[0, 1] \times [0, 1]$ box).

Through this transformation we show that the real weights used in the computation (p^*) are exactly the desired ones (p). In fact, in the following theorem we show that, in order to obtain results in which the vector $p^* = p$, the slope of the direction u must be fixed and the angle formed with the k axis must be constant.

Theorem 1. *Let $p^* \in \mathbb{R}^k$ be the vector used in the computation of the projections and $p \in \mathbb{R}^k$ the vector of weights with equal components. Thus:*

$$p^* = p \iff \text{the angle } \alpha_i \text{ formed by } u \text{ and the axis } i \in \{1, \dots, k\} \text{ is constant.}$$

Proof. *We recall that the ordering procedure is based on the projection of a vector in the direction u, that is considering the scalar product $< \hat{x}_i, u >$. So, in this case, the vector $p^* = u$. As a consequence, in order to prove the theorem, it is enough to show that $u = p$, that is $u_i = u_j$, $\forall i, j \in \{1, \dots, k\}$. Let us consider the vector $e_i = (0, \dots, 1, \dots, 0) \in \mathbb{R}^k$ for $i \in \{1, \dots, k\}$, the vector in the canonical basis of $\mathbb{R}^n$. It is clear that $< e_i, u > = u_i = \| u \| \cdot \| e_i \| \cos \alpha_i = \cos \alpha_i$. Thus $u_i = u_j$ for $i, j \in \{1, \dots, k\} \iff \alpha_i = \alpha_j$.*

Theorem 1 shows that, in order to have parameters with the same weight, it is necessary to apply the procedure of normalization proposed in this paper. In fact, in this case, the slope of u is fixed and the angle formed with the axis is constant. It is possible with this approach to consider also the case in which parameters do not have the same importance and so the weights are not constant. In this case, it is enough to multiply each component of the vector u with the corresponding weight in the vector p. In general, if the angle formed by the vector u and the axes $i \in \{1, \dots, k\}$ decreases, then the importance of that weight increases since the function $\cos x$ is decreasing for $x \in [0, \pi]$.

[3] Certainly this is not the only available normalizing procedure, but others do not modify the conclusions on the negative effect of the different *rv* of the variable to synthetize. The main problem is the need to homogenize, in a suitable way, data before they are calculated.

The problem of variables with different *rv* has an important application in the case of financial data. In fact, in order to evaluate different assets, financial agents have to carry out multidimensional ordering of data with parameters in very different scales, and the results can be strictly influenced by this circumstance.

In the next section we will present an application of such a problem to financial markets in order to avoid undesired effects in the evaluation process for decision makers.

4 Application to Financial Markets

In this section we present an application of this procedure to decision making in financial markets in the case in which, for example, the investor makes a decision taking into account two different financial indicators for each asset: CAP (in billions of euros) and EPS (in euros). Let us suppose that the decision maker gives each variable the same importance.

Considering data from the Italian Stock Market, in particular from S&PMIB, we consider 38 assets and data are updated to october 2006 (see Table 1).

Each asset is represented by a vector in a two dimensional space. In Fig. 1a, we present the cloud of data, on the x axis we have the EPS value and on the y axis the

Table 1. Assets from S&PMIB Italian Stock Market.

Asset	EPS	CAP	Asset	EPS	CAP
Aem	0.17	3.86	Italcementi	1.70	3.54
Autogrill	0.67	3.21	Lottomatica	1.31	4.47
Alleanza Ass.	0.53	7.81	Luxottica	1.13	10.65
Autostrade	1.26	13.39	Mediobanca	0.97	14.00
Alitalia	0.05	1.13	Mediolanum	0.36	4.24
Bca Intesa	0.44	31.19	Mondadori Ed.	0.51	1.89
Bca MPS	0.35	11.75	Mediaset	0.61	10.01
Bca PopIta	0.52	6.56	Pirelli	0.07	3.55
Bce Pop Ve/No	1.96	8.22	Seat PG	0.03	3.20
Bulgari	0.52	3.01	Parmalat	0.11	4.64
Capitalia	0.52	16.92	Bca PopMi	0.85	4.31
Enel	0.48	44.47	Ras Hold.	1.42	14.46
Eni	2.70	93.65	SanPaolo Imi	1.22	26.57
Gr. Ed. L'Espr	0.25	1.75	Saipem	1.05	7.60
Fiat	1.02	13.65	Snam ReteGas	0.24	7.47
Finmeccanica	1.25	7.51	Stmicroelettr.	0.84	12.50
Fondiaria SAI	2.90	4.64	Telecom	0.18	29.88
Fastweb	0.55	2.85	Terna	0.15	4.59
Generali	2.14	37.75	Unicredito	0.55	68.36

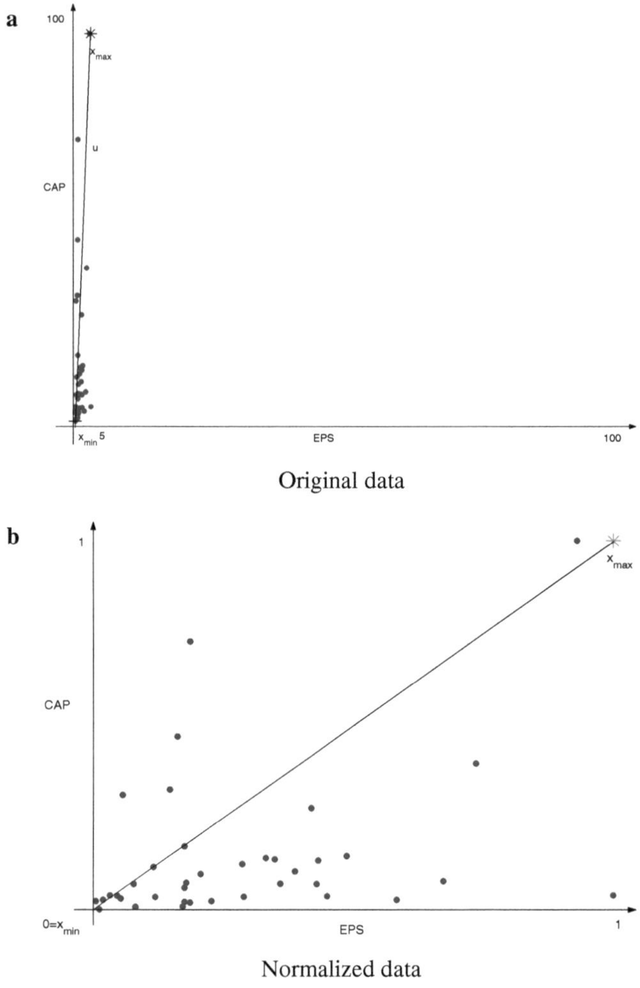

Fig. 1. In **a**, data are plotted in the original scale and it is evident how the parameter CAP is the most relevant. In **b**, due to the normalization procedure, the two variables are now in the same *rv* and as a consequence they have the same importance.

CAP. The *rv* is very different for the two parameters, in particular for EPS $rv = 2.85$ while for CAP $rv = 92.52$. The points x_{max} and x_{min} are obtained considering the maximum and minimum values of the coordinates in the data set respectively and so $x_{max} = (2.9, 93.65)$ and $x_{min} = (0.05, 1.13)$.

We apply the technique described in the previous section by (1), (2), and (3) to this data set calculating the projections on the vector u and considering the values in this direction for the ordering procedure. Since the vector p has equal weights, it is not necessary to modify the vector u, on the other hand, in the case in which the decision maker uses different weights for the variables, it is enough to multiply each component of the vector u with the corresponding weight.

Table 2. We present the two ordering procedures. Rank1 is the ranking obtained using data in original scale and Rank2 in the case of normalized data. The most relevant differences have been underlined in bold type.

Assets	Rank1	Rank2	Assets	Rank1	Rank2
Aem	29	34	Gr. Ed. L'Espr.	37	32
Alitalia	38	38	**Italcementi**	30	9
Alleanza Ass.	18	23	**Lottomatica**	26	13
Autogrill	32	24	Luxottica	15	12
Autostrade	12	10	Mediaset	16	22
Bca Intesa	5	15	Mediobanca	10	16
Bca MPS	14	26	Mediolanum	28	30
Bca PopIta	22	25	**Mondadori Ed.**	36	29
Bca PopMi	27	21	Parmalat	24	35
Bce Pop Ve/No	17	5	Pirelli	31	36
Bulgari	34	28	Ras Hold.	9	7
Capitalia	8	20	Saipem	19	17
Enel	3	8	SanPaolo Imi	7	6
Eni	1	1	Seat PG	33	37
Fastweb	35	27	**Snam ReteGas**	21	31
Fiat	11	14	Stmicroelettr.	13	18
Finmeccanica	20	11	**Telecom**	6	19
Fondiaria SAI	23	3	Terna	25	33
Generali	4	2	Unicredito	2	4

In order to show how the ordering procedure is influenced by the *rv* in the original data, we apply the modified ordering procedure proposed in this paper. In particular, we now consider the preventive normalization (4) to homogenize data preserving the original structure.

As shown in Fig. 1b, the data cloud has been considerably affected by the normalization procedure. As a result, the two variables are now in the same *rv* and, as a consequence of Theorem 1, it holds that $p^* = p$.

In Table 2 we present the ordering procedure obtained through original data (Rank1) and through normalized data (Rank2).

The results obtained show the importance of considering variables in the same *rv* very clearly. The ordering procedure produces very different results in particular for assets that have bad performance in CAP, that is the overweighted variable.

In the case of data in the original scale, in this simulation, the vector p^* used in the ordering procedure is very different from the desired one, in fact if we calculate the slope of the vector u in the case of data in the original scale we have $\alpha_1 = 88^o$ and $\alpha_2 = 2^o$ and so $p^* = (0.03, 0.97)$. As a result, the variable CAP has a weight of about 97% with respect to the other variable. Instead, in our procedure $\alpha_1 = \alpha_2 = 45^o$ and so $p^* = p$.

Let us consider, for instance, the case of the asset called Fondiaria SAI, whose EPS is equal to 2.9 (the maximum value) but its CAP is 4.6 (low). In the ordering procedure with original data, its ranking is very bad (position 23). On the other hand, applying the procedure proposed in this paper, the problem is resolved and the asset Fondiaria SAI rises to position 3. In this example, there are many other assets which are considerably affected by this feature. See for instance assets Bce Pop Ve/No, Enel, Italcementi, Finmeccanica, Bca Intesa, Mondadori Ed. and Snam ReteGas.

One possible way to quantify this feature is considering the Spearman cograduation coefficient calculated on the two rankings. In this case, the result is $\rho = 0.714$ and it confirms how, at least for some assets, the differences in the two rankings are considerable.

5 Conclusions

In this paper we have proposed a procedure for resolving the problem of defining complete ordering on multidimensional Euclidean spaces. In particular, we consider an ordering procedure in which data are projected in a "meaningful direction" in order to switch from a multidimensional problem to an unidimensional one. We have proposed a procedure in which results are not dependent on the scale of the data set considered. We show that if data are obtained considering variables in different scales, this can lead to results in which the *rv* of data affects the ordering procedure. In particular, data with a large *rv* are "overweighted" with respect to data with a small one. To resolve this problem we have proposed a normalizing procedure for the data before applying the ordering procedure. As a result, the *rv* of parameters is the same for all data, so the direction in which the projections are performed has a fixed slope.

We consider a real application of our proposal to decision making in financial markets in the case in which the investor makes a decision taking into due account two different financial indicators for each asset: CAP (in billions of euros) and EPS (in euros). The results show the importance of using the procedure proposed in this paper in order to obtain results that are consistent with the decision maker's tasks.

References

[LR04] Le Roux, B., Rouanet, H.: Geometric Data Analysis. Kluwer Academic Publishers, Dordrecht (2004)

[Del95] Delvecchio, F.: Scale di Misura e Indicatori Sociali. Cacucci Editore, Bari (1995)

[DR04] D'Esposito, M. R., Ragozini, G.: Atti della XLII riunione scientifica della Società Italiana di Statistica, Cleup, Padova, 51–54 (2004)

[Bar76] Barnett, V.: The Ordering of Multivariate Data. Journal of Royal Statistical Society, Ser. A, **139**, 318–354 (1976)

[RS05] Russo, M. A., Spada, A: Problemi statistici nella determinazione dell'indice di qualità (IGQ) per le cultivar di frumento duro. Atti del XXXVI Convegno della Società Italiana di Agronomia, 29–30 (2005)

[RS06] Russo, M. A., Spada, A.: Determining Durum Wheat Quality to Obtain EU Incentives: Statistical Problems and Tentative Solution. Atti della XLIII Riunione Scientifica della Società Italiana di Statistica, 367–370 (2006)

[CRZ99] Corbellini, A, Riani, M., Zani, S.: New Methods for Ordering Multivariate Data: an Application to the Performance of Investments Funds. Applied Stochastic Models in Business and Industry, 485–493 (1999)

[Sci98] Scippacercola S.: Interpoint Distances in the Multivariate Data Ordering, Statistica Applicata, vol.**10**:**1**, 73–83 (1998)

A Biometric Risks Analysis in Long Term Care Insurance

Susanna Levantesi and Massimiliano Menzietti

Summary. This paper deals with problem of analyzing uncertainty arising from both mortality and disability risks in Long Term Care (LTC) covers. To this purpose some projected demographical scenarios are developed. A biometric risks analysis is performed at portfolio level according to both loss function and risk reserve. The probabilistic structure adopted is consistent with multiple state models, based on a time-continuous Markov chain.

Key words: Long Term Care Covers; Biometric risks; Demographical trends; Loss function; Risk reserve.

1 Introduction

The purpose of LTC insurance is to offer protection against costs arising from future deterioration in health. The most critical issues in defining LTC risk come from lack of reliable experience data, uncertainty about future trends in mortality and disability amongst the elderly, adverse selection and moral hazard.

This paper aims to analyze biometric risks deriving from disability and lifetime duration. It is structured as follows. In Section 2, a multiple state model for LTC covers is defined. Section 3 deals with the estimate of transition intensities necessary for pricing and reserving, while Section 4 deals with the construction of a set of projected demographical scenarios. Insurance benefits, premiums and reserves are described in Section 5, while a biometric risk analysis according to loss function and risk reserve is described in Section 6. Finally, some results are reported and discussed in Section 7 as well as some concluding remarks.

2 Multiple State Model

LTC insurance is usually modelled according to a multiple state model (an extensive description of such models can be found in [HP99]).

Let $S(t)$ represent the random state occupied by the insured at time t, for any $t \geq 0$, where t represents the policy duration and 0 is the time of entry. The possible

Cira Perna, Marilena Sibillo (Eds.): Mathematical and Statistical Methods in Insurance and Finance

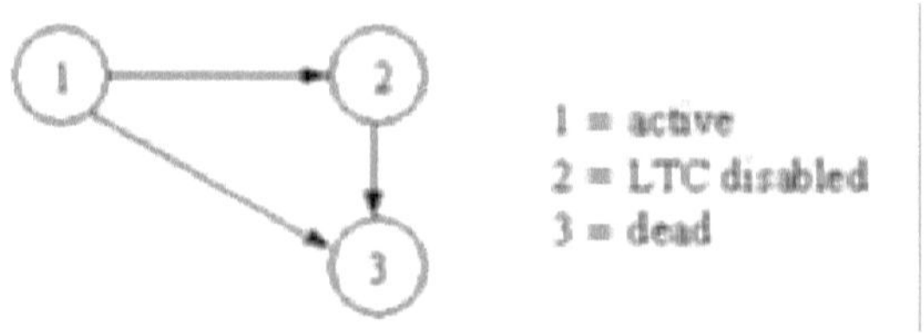

Fig. 1. A multiple state model for LTC insurance with one level of disability.

realizations of $S(t)$ are: 1 = "active" (or healthy), 2 = "LTC disabled", 3 = "dead". Note that only one level of disability is considered.

We assume $S(0) = 1$. Moreover, because of the usually chronic character of disability, the possibility of recovery from the LTC state is disregarded; therefore transition $2 \rightarrow 1$ is ignored.

The graph in Fig. 1 illustrates the states occupied by the policyholder in a time instant and the possible direct transitions between states.

Let us assume that the stochastic process $\{S(t);\ t \geq 0\}$ is a time-continuous, time inhomogeneous, three state Markov chain.

Let us define transition probabilities:

$$P_{ij}(t, u) = \Pr\{S(u) = j \ |S(t) = i\} \qquad 0 \leq t \leq u, \qquad i,j \in \{1, 2, 3\}\ , \tag{1}$$

and transition intensities:

$$\mu_{ij}(t) = \lim_{u \to t} \frac{P_{ij}(t, u)}{u - t} \qquad t \geq 0, \quad i,j \in \{1, 2, 3\},\ \ i \neq j\ . \tag{2}$$

According to the model depicted in Fig. 1, the following transition probabilities are used in premium and reserve calculation:

$$P_{11}(t, u) = \exp\left(-\int_t^u \mu_{12}(s) + \mu_{13}(s)\, ds\right) \tag{3}$$

$$P_{22}(t, u) = \exp\left(-\int_t^u \mu_{23}(s)\, ds\right) \tag{4}$$

$$P_{12}(t, u) = \int_t^u P_{11}(t, s) \cdot \mu_{12}(s) \cdot P_{22}(s, u)\, ds \tag{5}$$

$$P_{13}(t, u) = \int_t^u P_{11}(t, s) \cdot \mu_{13}(s) + P_{12}(t, s) \ \cdot \mu_{23}(s) ds \tag{6}$$

$$P_{23}(t, u) = \int_t^u P_{22}(t, s) \cdot \mu_{23}(s) ds\ . \tag{7}$$

3 Estimation of Transition Intensities

In Italy, LTC covers are recent products and insurance experience data about LTC claims are still inadequate for pricing and reserving, hence population data collected from Italian national source represent, at that moment, the best available data. Italian Life-Table (SIM 1999) and statistical data about people reporting disability (see [Ist00]) have been used in numerical implementation.

Probabilities $P_{11}(t, u)$, $P_{12}(t, u)$ and $P_{13}(t, u)$ have been calculated for each integer (t, u) starting from ISTAT data. Regarding LTC disabled mortality, we assume that it is related to active mortality (since specific data are lacking, such hypothesis seems quite reasonable):

$$P_{23}(t, t+1) = k(t) \cdot P_{13}(t, t+1)\,, \tag{8}$$

where the time-dependent factor values, $k(t)$, come from experience data from an important reinsurance company.

Let us consider $x = 50$ as age at policy issue. Results show that actives' mortality can be well approximated by the Weibull law, while transition intensities $\mu_{12}(t)$ show an exponential behaviour suggesting the use of the Gompertz law:

$$\mu_{13}(t) = \frac{\beta}{\alpha}\left(\frac{x+t}{\alpha}\right)^{\beta-1} \qquad \alpha, \beta > 0\,, \tag{9}$$

$$\mu_{12}(t) = \eta \cdot e^{\lambda(x+t)} \qquad \eta, \lambda > 0\,, \tag{10}$$

$\mu_{23}(t)$ is expressed in terms of $\mu_{13}(t)$ according to the time-dependent coefficient $K(t)$, well approximated by the function $\exp(c_0 + c_1 t + c_2 t^2)$:

$$\mu_{23}(t) = K(t)\mu_{13}(t)\,. \tag{11}$$

4 Demographic Scenarios

When LTC annuity benefits are concerned, a key point in actuarial evaluations is to measure the risk coming from the uncertainty in disability duration, i.e. the time spent in LTC disabled state. To this purpose, it is necessary to make some assumptions about the link between mortality and disability. As regards to such link, three main theories have been formulated: pandemic, equilibrium and compression theory (for an overall review see [OP01]).

The uncertainty in future mortality and disability of the elderly suggests the need to adopt different projected scenarios in benefit evaluation. Therefore, some reliable scenarios have been defined according to these theories, including projection of both mortality and disability. Such method of defining demographical scenarios has been performed by Ferri and Olivieri [FO00].

Let us define the time expected to be spent in state j for a person in state i at time t:

$$\bar{e}_{ij}(t) = \int_t^{\infty} P_{ij}(t, u)du\,. \tag{12}$$

Table 1. Life expectancies at age 65.

S_i	$p(S_i)$	$e_{11}^{S_i}(0)$	$e_{12}^{S_i}(0)$	$e_{22}^{S_i}(0)$	$\frac{\bar{e}_{11}^{S_i}(0)}{\bar{e}_{1.}^{S_i}(0)}$	$\frac{\bar{e}_{11}^{S_i}(0)}{\bar{e}_{11}^{S_B}(0)} - 1$	$\frac{\bar{e}_{12}^{S_i}(0)}{\bar{e}_{12}^{S_B}(0)} - 1$	$\frac{\bar{e}_{1.}^{S_i}(0)}{\bar{e}_{1.}^{S_B}(0)} - 1$
S_B	–	14.6746	1.4253	5.8155	91.15%	–	–	–
S_{1a}	4%	17.0550	1.3247	7.1766	92.79%	16.22%	−7.06%	14.16%
S_{1b}	12%	15.9807	1.7599	7.1766	90.08%	8.90%	23.48%	10.19%
S_{1c}	4%	15.0898	2.1151	7.1766	87.71%	2.83%	48.40%	6.86%
S_{2a}	12%	17.7646	1.5553	8.1609	91.95%	21.06%	9.12%	20.00%
S_{2b}	36%	16.5839	2.0560	8.1609	88.97%	13.01%	44.25%	15.78%
S_{2c}	12%	15.6143	2.4618	8.1609	86.38%	6.40%	72.72%	12.27%
S_{3a}	4%	18.4757	1.8270	9.2701	91.00%	25.90%	28.19%	26.11%
S_{3b}	12%	17.1808	2.4022	9.2701	87.73%	17.08%	68.54%	21.63%
S_{3c}	4%	16.1282	2.8651	9.2701	84.92%	9.91%	101.02%	17.97%

In order to evaluate the time spent in state 1 and 2 we calculate $\bar{e}_{11}(t)$, $\bar{e}_{12}(t)$ and $\bar{e}_{22}(t)$. Since only one cohort of policyholders with same age at policy issue and same year of entry has been considered, the calendar year has been omitted in the life expectancy notation. Note that total life expectancy for a healthy person is expressed as: $\bar{e}_{1.}(t) = \bar{e}_{11}(t) + \bar{e}_{12}(t)$.

The theories mentioned above can be expressed in terms of life expectancies:

- Compression: $\bar{e}_{1.}(t)$ rises with time increase with a major contribution (in relative terms) from $\bar{e}_{11}(t)$.
- Equilibrium: both $\bar{e}_{11}(t)$ and $\bar{e}_{12}(t)$ rise with time increase, at similar rates.
- Pandemic: $\bar{e}_{1.}(t)$ rises with time increase, with a major contribution (in relative terms) from $\bar{e}_{12}(t)$.

The basic scenario, S_B, defined according to ISTAT data, has been considered as a starting point to construct projected scenarios. Projected mortality has been developed according to ISTAT projections (low, main and high hypothesis, see [Ist02]), where total life expectancy increases anyhow. In order to represent compression, equilibrium and pandemic theories, three different sets of parameters for $\mu_{12}(t)$ have been defined starting from the basic scenario. Combining mortality and disability projections, nine scenarios have been obtained (see Table 1).

Risk analysis is performed in terms of both scenario and stochastic approach. This latter approach considers each scenario, S_i, as a possible result, according to a given probability, $p(S_i)$.

5 Benefits, Premiums, and Reserve

Because LTC covers are frequently offered as complementary with life insurance, we consider an insurance policy which provides the following benefits:

1. A deferred annuity paid at an annual rate $b_1(t)$, when the insured is healthy.

2. A deferred enhanced annuity paid at an annual rate $b_2(t) > b_1(t)$, when the insured is LTC disabled.
3. A lump sum benefit $c_3(t)$, if death occurs before the deferment period.

All benefits are supposed to be constant with time. Let n be the deferment period, the actuarial value at time 0 of these benefits, $\Pi(0,\omega)$ is given by the following:

$$\begin{aligned}\Pi(0,\omega) &= b_1 \cdot P_{11}(0,n)v^n a_{11}(n,\omega) + b_2[P_{11}(0,n)a_{12}(n,\omega)\\ &\quad + P_{12}(0,n)a_{22}(n,\omega)]v^n + c_3[A_{113}(0,n) + A_{123}(0,n)]\,,\end{aligned} \tag{13}$$

where: $a_{ij}(t,u) = \sum_{s=t}^{u-t-1} P_{ij}(t,s) \cdot v^{s-t} \quad \forall i,j \in 1,2,$

and $A_{ijk}(t,u) = \sum_{s=t}^{u-t-1} P_{ij}(t,s) \cdot P_{jk}(s,s+1) \cdot v^{s+1-t} \qquad \forall i,j \in 1,2 \text{ and } k = 3.$

Reserve for active lives is defined as:

$$V_1(t) = \begin{cases} \Pi(t,\omega) - \pi \cdot a_{11}(t,n) & t < n \\ b_1 a_{11}(t,\omega) + b_2 a_{12}(t,\omega) & t \geq n\,, \end{cases} \tag{14}$$

where $\pi = \frac{\Pi(0,\omega)}{a_{11}(0,n)}$ represents the annual constant premium, paid if the insured is active. Reserve for LTC disabled is given by:

$$V_2(t) = \begin{cases} b_2 \cdot P_{22}(t,n)v^{n-t}a_{22}(n,\omega) + c_3 A_{223}(t,n) & t < n \\ b_2 \cdot a_{22}(t,\omega) & t \geq n\,. \end{cases} \tag{15}$$

Let α, β and γ be the premium loading for acquisition, premium earned and general expenses, respectively. Gross premium is defined as:

$$\pi^T = \frac{\pi}{1 - \frac{\alpha}{a_{11}(0,n)} - \beta - \gamma\,\frac{a_{11}(0,\omega)+a_{12}(0,\omega)}{a_{11}(0,n)}}\,. \tag{16}$$

6 Risk Analysis

In order to analyze risk arising from uncertainty in mortality and disability trends, we firstly consider the loss function at portfolio level, $L(t)$, defined as the difference between the random present value of future benefits, $Y(t)$, and the random present value of future premiums, $X(t)$.

Let I_E be the event E indicator; $Y(t)$ at time $t = 0,1,2,\ldots$ is given by:

$$Y(t) = \sum_{j=1}^{N(t)} \left[\sum_{h=t+1}^{n} v^{h-t} c_3 I_{\{S_j(h)=3\}} + \sum_{h=n}^{\omega} v^{h-t} \left(b_1 I_{\{S_j(h)=1\}} + b_2 I_{\{S_j(h)=2\}} \right) \right], \tag{17}$$

where $N(t)$ is the number of contracts in force at time t and $S_j(h)$ the state occupied by the j-th policy at time h. $X(t)$ at time t is:

$$X(t) = \sum_{j=1}^{N(t)} \sum_{h=t}^{n} v^{h-t} \pi\, I_{\{S_j(h)=1\}}\,. \tag{18}$$

As risk measure at time t, we take into account the following risk index:

$$r(t) = \frac{\sqrt{Var[L(t)]}}{E[Y(t)]}. \tag{19}$$

Now, as alternative risk measure we consider the risk reserve which involves expenses, investment, and capital provision. According to a well known formula of risk theory (see [DPP94]), the risk reserve at the end of year t is defined as:

$$U(t) = U(t-1) + P^T(t) + J(t) - E(t) - B(t) - \Delta V(t) + K(t), \tag{20}$$

where $P^T(t)$ is the gross premiums income, $J(t)$ the investment returns on assets, $E(t)$ the expenses, $B(t)$ the outcome for benefits, $\Delta V(t)$ the increment in reserve and $K(t)$ the capital flows.

7 Portfolio Simulation Results

Let us consider one cohort of policyholders with the same age at policy issue, $x = 50$, same year of entry (time 0), same benefit amounts and same risk class. Let us assume: $n = 15$; $\omega = 60$; $b_1 = 100$; $b_2 = 200$; $c_3 = 1000$; $\alpha = 60\%$; $\beta = 2\%$; $\gamma = 8\%$. In order to perform a biometric risks analysis, a deterministic financial structure is adopted with i=3%.

In Table 2, $\sqrt{Var[L(t)]}$, $E[Y(t)]$ and $r(t)$ values at time 0 under both scenario and stochastic approach are shown for the insurance policy designed. Risk analysis is also performed for each single benefit considered according to scenario approach (see Table 3).

Under scenario approach, results show high values of $r(0)$ for term life insurance and LTC, compared with life annuity. Note that "full" policy riskiness is less than its components. Furthermore, the risk of mortality random fluctuations can be greatly lowered by increasing the portfolio size for the diversification effect. On the other hand, under stochastic approach, $r(0)$ decreases more slowly compared to scenario approach since the model incorporates the risk of systematic deviations.

Table 2. Results for the "full" policy.

	Scenario approach			Stochastic approach		
$N(0)$	$E[Y(0)]$	$\sqrt{Var[L(0)]}$	$r(0)$	$E[Y(0)]$	$\sqrt{Var[L(0)]}$	$r(0)$
100	96,955	3,852	0.0397	97,000	4,689	0.0483
200	193,905	5,480	0.0282	194,020	7,510	0.0387
500	484,758	8,676	0.0179	484,852	15,948	0.0329
1000	969,507	12,000	0.0124	969,228	28,941	0.0298
5000	4,847,577	27,458	0.0057	4,847,024	137,033	0.0283

Table 3. $r(0)$ for each policy component - Scenario approach.

	Life annuity	LTC annuity	Term life
$N(0)$	$b_1 = b_2; c_3 = 0$	$b_1 = c_3 = 0$	$b_1 = b_2 = 0$
100	0.0441	0.1956	0.3634
200	0.0308	0.1389	0.2547
500	0.0192	0.0880	0.1592
1000	0.0135	0.0605	0.1100
5000	0.0057	0.0275	0.0519

The risk analysis according to risk reserve is performed for stochastic approach assuming that $K(0) = 0$ and $U(0) = 0$. Premiums are calculated according to central scenario; if real scenario is different from the central one, we assume that the insurance company adjusts reserves with a 5 year delay.

If we introduce a safety loading on demographical bases (a 10% reduction of death probabilities, π^T=106.53), risk reserve translates towards the top, reducing portfolio riskiness (see Fig. 3).

Simulation results are reported in terms of ratio between risk reserve and reserve. Fig. 2 (left) shows results for a portfolio of 1,000 insureds, assuming no safety

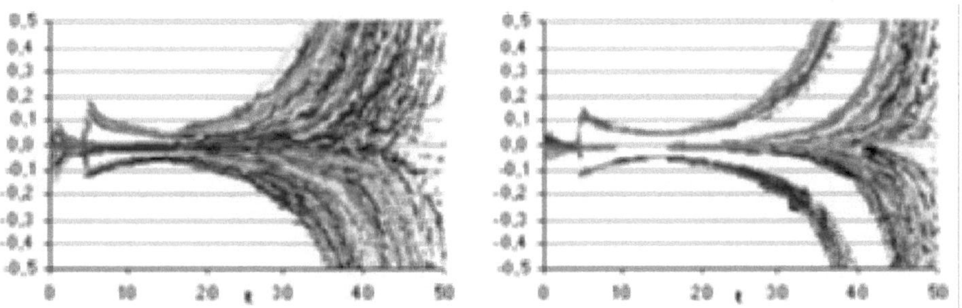

Fig. 2. $U(t)/V(t)$ for 1,000 and 10,000 insured, respectively.

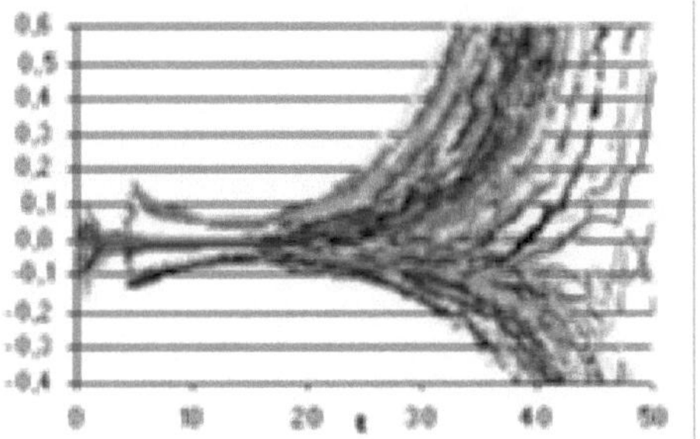

Fig. 3. $U(t)/V(t)$ for 1,000 insured (with safety loading).

loading (π^T=102.59): only three branches of trajectories instead of nine (i.e. the scenarios considered) are observable due to the strong weight of mortality with respect to disability. By increasing the number of insured, the risk of random fluctuations decreases (it is a pooling risk), while systematic deviations risk remains unchanged (it is a non-pooling risk) (see Fig. 2 (right)). Our analysis shows high systematic deviations risk in LTC cover, that can be partially reduced by combining LTC with other insurance covers and including a safety loading. Nonetheless, in order to assure solvency requirements, the insurance company should involve reinsurance strategies and appropriate capital flows policy. These arguments will be the subject of future works together with a financial risk analysis, assuming an ongoing concern approach.

References

[DPP94] Daykin, C. D., Pentikäinen, T., Pesonen, M.: Practical Risk Theory for Actuaries. Chapman & Hall, London (1994)

[FO00] Ferri, S., Olivieri, A.: Technical bases for LTC covers including mortality and disability projections. In: Proceedings of the 31st International ASTIN Colloquium, Porto Cervo (Italy) (2000)

[HP99] Haberman, S., Pitacco, E.: Actuarial Models for Disability Insurance. Chapman & Hall, London (1999)

[Ist00] ISTAT: Le condizioni di salute e ricorso ai servizi sanitari. Settore Famiglie e societa', ISTAT, Roma (2000)

[Ist02] ISTAT: Previsioni della popolazione residente per sesso età e regione dal 1.1.2001 al 1.1.2051. Settore Popolazione, ISTAT, Roma (2002)

[OP01] Olivieri, A., Pitacco, E.: Facing LTC risks. In: Proceedings of the 32nd International ASTIN Colloquium, Washington (2001)

Clustering Financial Data for Mutual Fund Management

Francesco Lisi and Marco Corazza

Summary. In this paper, an analysis of the performances of an active and quantitative fund management strategy is presented. The strategy consists of working with a portfolio constituted by 30 equally-weighted stock assets selected from a basket of 397 stock assets belonging to the Euro area. The asset allocation is performed in two phases: in the first phase, the 397 stock assets are split into 5 groups; in the second, 6 stock assets are selected from each of the group. The analysis focuses: i) on the specification of quantitative approaches able to effect the group formation; ii) on the definition of a profitable active and quantitative fund management strategy; iii) on the quantitative investigation of the contribution individually provided by each of the two phases to the total profitability of the fund management strategy.

Key words: Mutual fund management; Euro stoxx 50 index; Relative strength index; Cluster analysis; Autocorrelation structure; GARCH models.
J.E.L. classification number: G11, C14.
M.S. classification number: 91B28, 91C20, 62M10.

1 Introduction

A mutual fund can be thought of as a kind of collective cash in which the savings of a plurality of investors come together so that they are profitably invested in suitable assets by an *ad hoc* established and regulated company.

The Italian-law mutual funds are required to quote a parameter, called benchmark, which indicates their risk-expected return profile. Usually, the benchmark is given by a financial index. The comparison between the performances of a fund and the performances of the related benchmark is a way of evaluating the fund management.

In general, it is possible to classify fund management styles in (at least) two ways: active *versus* passive management, and discretionary *versus* quantitative management. One speaks of active/passive management when the target of the fund manager consists in exceeding/replicating the performances of the chosen benchmark. In both these cases, the dynamic asset allocation process can be performed by resort-

Cira Perna, Marilena Sibillo (Eds.): Mathematical and Statistical Methods in Insurance and Finance

ing to the expertise of human managers (*i.e.* to the discretionary management), or to quantitative methodologies (*i.e.* to the quantitative management).

In this paper, an analysis of the performances of an active and quantitative fund management strategy is presented. The strategy starts from the practice, followed by an Italian financial advice company, of working with a portfolio constituted by 30 equally-weighted stock assets selected from a basket of 397 traded in financial markets of the Euro area.

The asset allocation is performed in two phases: first, the 397 stock assets are split into 5 groups; then, from each group, 6 stock assets are selected.

Initially, group formation was carried out by a discretionary-based approach in order to obtain homogeneous groups. In particular, this grouping was performed on the basis of the belonging of stock assets to various economic sectors and on the basis of the related growth prospects. The group formation is updated every six months.

The stock asset selection from each of the 5 groups is based on a variant of a well known tool of technical analysis, called relative strength, given by

$$RS_{i,t} = P^*_{i,t} \,/ I^*_t \;, \; i = 1, \ldots, 397, \; t = 1, \ldots, T,$$

where T is the number of the considered time periods, $P^*_{i,t} = 100(P_{i,t}/P_{i,0})$, with $P_{i,t}$ price of the i-th stock asset at time t, and $I^*_t = 100(I_t/I_0)$, with I_t value of the chosen benchmark at time t (in this case I_t is the Euro stoxx 50 financial index). At first, the stock assets within each group are sorted in decreasing order with respect to the value of the so-called relative strength index (RSI), given by

$$RSI_{j,t} = MM^j_{S,t} \Big/ MM^j_{L,t} \;, \; j = 1, \ldots, N_g, \; g = 1, \ldots, 5,$$

where N_g is the number of stock assets belonging to the g-th group, $MM^j_{S,t} = \sum_{s=0}^{K_S-1} RS_{j,t-s}/K_S$ and $MM^j_{L,t} = \sum_{l=0}^{K_L-1} RS_{j,t-l}/K_L$ are, respectively, the "short" and the "long" moving average of $RS_{j,t}$, with $K_S < K_B$ (see, for details, [K05]).

Then, from each of the so-ordered groups, the first 6 stock assets are selected.

The purpose of this stock asset selection procedure is typical of the technical analysis, and consists of detecting those stock assets whose prices are recently quickly increased with respect to the chosen benchmark.

The ordering of one of the six groups, and consequently the portfolio composition, is updated every week.

Within this operative framework, the analysis focuses:

- On the specification of approaches able to effect the group formation in a quantitative way instead of in a discretionary one.
- On the definition of an active and quantitative fund management strategy whose performance are, on average, better than the ones of the chosen benchmark.
- On the quantitative investigation of the contribution individually provided by the grouping phase and by the stock asset selection phases to the total profitability of the fund management strategy.

2 Clustering Financial Data

In order to reduce the subjectivity in the grouping phase, the discretionary-based approach is replaced with a quantitative one based on cluster analysis (see, for instance, [M79]).

It is well known that cluster analysis aims to classify N statistical units, with respect to P variables, in a small number of groups called clusters. In particular, the clustering is carried out in such a way as to guarantee the highest "similarity" within groups and the highest "dissimilarity" between groups.

The cluster analysis starts by considering the so-called data matrix $\mathbf{X} = [x_{i,j}]$, with $i = 1, \ldots, N$ and $j = 1, \ldots, P$, where $x_{i,j}$ is the value assumed by the j-th variable for the i-th statistical unit. So, the data matrix contains all the relevant information necessary for clustering.

In order to measure the similarity/dissimilarity between statistical units, a similarity/dissimilarity measure is utilized. Usually, this is specified by a distance function $d\,(\cdot,\cdot)$: $\mathbb{M}^2 \to \mathbb{R}_+$, where $\mathbb{M}$ is the space of the modalities. This function allows the determination of the so-called distance matrix $\mathbf{D} = [d_{i,j}]$, with $i, j = 1, \ldots, N$, where $d_{i,j}$ is the distance between the i-th statistical unit and the j-th one.

Clustering algorithms applicable to matrix $\mathbf{D}$ can be of various kinds. In this paper the Ward's iterative aggregation method is utilized. In short, this method aggregates at each iteration, those two clusters, among the existing ones, in correspondence of which the increase of the deviance of the cluster in making is minimized (see, for more details, [H01]).

Of course, in applying cluster analysis, the choices of the variables and the distance function are critical. In this paper, three variable/distance function proposals are taken into account.

The first proposal follows the approach suggested in [G00]. It considers the patterns of stock asset returns. In particular, it starts by considering the data matrix

$$\mathbf{X} = \left[r_{i,t}\right],\ i = 1, \ldots, 397,\ t = 1, \ldots, T,$$

where $r_{i,t}$ is the logarithmic return of the i-th stock asset at time t. As far as $d(\cdot,\cdot)$ concerns, the Euclidean metric is considered.

The second proposal focuses on the autocorrelation structures of the time series of both stock asset returns and of their squares. The related data matrix is given by

$$\mathbf{X} = \left[\rho_k(\underline{r}_i) \mid \phi_k(\underline{r}_i) \mid \rho_k(\underline{r}_i^2) \mid \phi_k(\underline{r}_i^2)\right],\ i = 1, \ldots, 397,\ k = 1, \ldots, 15,$$

where $\underline{r}_i$ is the time series of the i-th stock asset returns, and $\rho_k(\cdot)$ and $\phi_k(\cdot)$ are, respectively, the global and the partial autocorrelation functions at lag k. Again, the Euclidean metric is considered.

The third proposal refers to the volatility of the stock asset returns, and groups are formed with respect to models describing the dynamics of the volatilities themselves. In particular, for each time series $\underline{r}_i$, with $i = 1, \ldots, 397$, a GARCH(1, 1) model is fitted under the assumption that the related conditional volatility, *i.e.* $\sigma_{i,t}^2$, can be described as $\sigma_{i,t}^2 = \omega_i + \alpha_i r_{i,t-1}^2 + \beta_i \sigma_{i,t-1}^2$ (see [B86] for details). After these models are fitted, the distance matrix **D** can be obtained – referring to parameters α_i and β_i – by considering the distance $d_{i,j}$ between GARCH models proposed in [O04], which is, in its turn, a generalization of the metric for ARMA models proposed in [P90],

$$d_{i,j} = \sqrt{\sum_{l=0}^{+\infty} \left(\alpha_i \beta_i^l - \alpha_j \beta_j^l\right)^2} = \cdots =$$

$$= \sqrt{\frac{\alpha_i^2}{1-\beta_i^2} + \frac{\alpha_j^2}{1-\beta_j^2} + \frac{2\alpha_i \alpha_j}{1-\beta_i \beta_j}}, \quad i, j = 1, \ldots, 397.$$

3 Applications

In this section, the three proposed strategies for fund management are applied and compared, with a benchmark and with some randomized strategies.

To this end, the daily close prices of 397 stock assets belonging to markets of the Euro area are considered from January 2002 to June 2005. The financial markets involved are Austrian, Belgian, Dutch, Finnish, French, German, Italian, Portuguese and Spanish. The stock assets belonging to the basket are those characterized in the 30 days before the basket formation by a mean trading volume of, at least, 1,500,000 Euro.

The performances of all the considered strategies and of the benchmark are evaluated out-of-sample along three semesters in succession (January 2004 to June 2004; July 2004 to December 2004; January 2005 to June 2005). Moreover, the data related to the previous 24 months of each semester are used to make the in-sample setting of the strategies themselves.

As far as the measurement of performances is concerned, the following quantities are taken into account: two profitability indicators, *i.e.* the period geometric return (PGR) and the mean quarterly geometric return ($MQGR$), and two risk indicators, *i.e.* the maximum percentage draw down ($MPDD$) and the mean pay-back time ($MPBT$).[1] It is to note that the performances of all the considered strategies are measured net of the costs of transaction, settlement, slippage etc. On the whole, these costs are set equal to 0.3% of the invested capital.

As reference benchmark, a portfolio composed of the 397 considered stock assets, all equally-weighted, is taken into account. This benchmark is preferred to the Euro stoxx 50 index because during the period of interest, the former performed better than the latter. Therefore, under the point of view of the comparison with the results of the strategies, this choice is unfavorable for the investigation performed

[1] In qualitative terms, the pay-back time indicates the length of time between the instant in which a minimum of the profitability of a considered asset occurs, and the instant in which the subsequent maximum of the same profitability occurs.

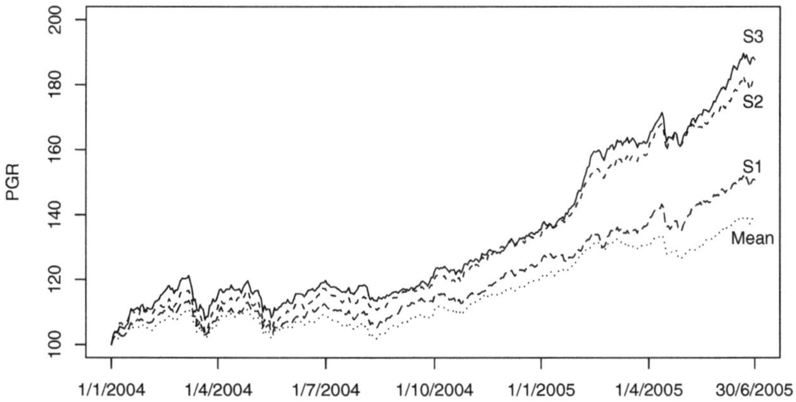

Fig. 1. Out-of-sample behavior of PGR for strategies and benchmark.

here. Moreover, it is also to note that the performances of the chosen benchmark are measured gross of the costs. So, this choice is once again unfavorable.

As first analysis, the three fund management strategies and the benchmark are considered. In the grouping phase, these strategies utilize, respectively, the proposal "patterns of returns/Euclidean metric" ($S1$), the proposal "autocorrelation structures/Euclidean metric" ($S2$), and the proposal "GARCH volatility modelling/distance between GARCH models" ($S3$). As far as the stock asset selection is concerned, all the strategies use the RSI-based approach.

The best in-sample performance was achieved by strategy $S3$ in all the investigated periods. Fig. 1 shows, for all the strategies, the out-of-sample behaviors of the related PGRs, which confirm the indications of the in-sample analysis. Table 1 shows that the values of the indicators PGR and $MQGR$ related to the three strategies are all greater than those of the benchmark (Mean). In particular:

- The values of both the indicators PGR and $MQGR$ for strategy $S3$ are about twice the corresponding indicators related to the benchmark.
- The more profitable strategy is $S3$, followed by $S2$, $S1$ and the benchmark.

With respect to the risk indicators, the value of the indicator $MPDD$ related to strategy $S3$ is the 37.55% greater than that of the corresponding indicator of the benchmark, whereas the values of the indicators $MPBT$ related to all strategies and

Table 1. Out-of-sample performance indicators.

Indicator	Benchmark	Strategy $S1$	Strategy $S2$	Strategy $S3$
PGR	39.33%	50.77%	81.54%	87.74%
$MQGR$	4.83%	6.13%	9.32%	9.47%
$MPDD$	8.07%	9.71%	11.03%	11.10%
$MPBT$	8.03 days	7.48 days	6.64 days	7.33 days

Fig. 2. Out-of-sample behavior of the differences between the PGR of strategy $S3$ and the PGR of strategy SR.

the benchmark are about the same. Synthetically, risk increases less proportionally than return increases.

In the second analysis, the best of the previous strategies, *i.e.* $S3$, is considered in order to determine the contribution provided by the grouping phase to the total profitability of the strategy. To this end, strategy $S3$ is compared with a semi-random strategy, named SR, in which the groups are determined completely at random. Instead, the selection phase is carried out for both the strategies by the same RSI-based approach. Since strategies $S3$ and SR differ only in the group formation, and since the random choice of the clusters is a way to "nullify" the effect of grouping, the (possible) differences in the respective performances should be attributed only to the way in which the clusters are determined. Moreover, in order to avoid any (possible) bias coming from the use of a single random grouping, the performances of the strategy SR are determined as the mean performance coming from 100 random clusterings.

Fig. 2 shows the out-of-sample behavior of the difference between the indicator PGR related to strategy $S3$ and that related to the strategy SR. This difference is almost always positive and has a maximum equal to 7.80%. In particular, its mean value is equal to 2.20%, which can be considered the contribution to the total profitability of the clustering phase of the "GARCH(1, 1) volatility modelling/distance between GARCH models" approach.

Now, assuming the additivity of the contributions of each of the two phases to the total profitability (in terms of PGR) of a fund management strategy, it is possible to determine the contribution provided by the stock asset selection phase by the difference between the total profitability of the strategy itself and the contribution of the grouping phase. Fig. 3 shows the out-of-sample behavior of this contribution. Not surprisingly, this contribution (which coincides with the total profitability of the strategy SR) is considerably greater than the contribution provided by the group formation phase.

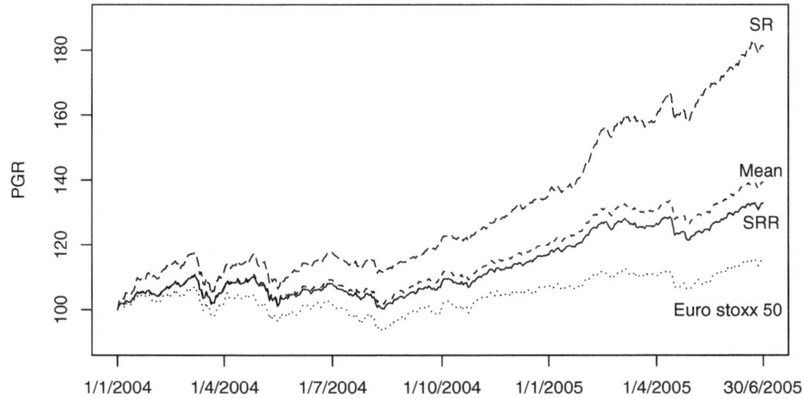

Fig. 3. Out-of-sample behavior of indicators PRG.

An alternative way to look at the contribution provided by the stock asset selection phase of strategy $S3$ consists of comparing it with a fully random strategy, named SRR, in which both clustering and stock selection happen in a completely random manner. The total profitability (in terms of PGR) of strategy SRR can be considered as the "zero-level" of the performances.[2] In fact, any positive but lower than zero-level profitability has to be attributed only to the joint effect of chance and of the market trend. Fig. 3 shows the out-of-sample behavior of the PGRs of strategies SR and SRR, of the benchmark, and of the Euro stoxx 50 index. It shows:

- That strategy SR is more profitable than the other considered asset allocation approaches.
- That the benchmark (Mean) performs better than strategy SRR, and that SRR outperforms the Euro stoxx 50 index. This is likely to be due to a small-cap effect. In fact, during the out-of-sample periods, small- and medium-capitalization firms performed better than high-capitalization ones.

4 Concluding Remarks

In this paper some alternative strategies for mutual fund management are proposed, compared among them, with a benchmark (Euro stoxx 50) and with some randomized stock asset selection procedures.

Results of these analyses (see Fig. 1 and Table 1) suggest that an active and quantitative approach to fund management, instead of a passive and discretionary ones, can systematically turn out profitable.

The results also suggest (see Fig. 3) that systematically choosing a financial index as benchmark could be misleading since its profitability may be positive only

[2] Also the profitability of strategy SRR is determined as mean of the performances from 100 simulations.

by chance. It leads to the proposal of using as benchmark, the best-performing one chosen between a reference financial index and a fully-random strategy.

References

[B86] Bollerslev, T.: Generalized autoregressive conditional heteroskedasticity. Journal of Econometrics, **31**, 307–327 (1986)

[G00] Gavrilov, M., Anguelov, D., Indyk, P., Motwani, R.: Mining the stock market: which measure is best?. In: Proceedings of the Sixth ACM SIGKDD International Conference on Knowledge Discovery and Data Mining. ACM Press, New York (2000)

[H01] Hastie, T., Tibshirani, R., Friedman, J.: The Elements of Statistical Learning. Data Mining, Inference, and Prediction. Springer, New York (2001)

[K05] Kaufman, P. J.: New Trading Systems and Methods. Wiley, New York (2005)

[M79] Mardia, K. V., Kent, J. T., Bibby, J. M.: Multivariate Analysis. Academic Press, London (1979)

[O04] Otranto, E.: Classifying the markets volatility with ARMA distance measures. Quaderni di Statistica, **6**, 1–19 (2004)

[P90] Piccolo, D.: A distance measure for classifying ARIMA models. Journal of Time Series Analysis, **11**, 153–164 (1990)

Modeling Ultra-High-Frequency Data: The S&P 500 Index Future

Marco Minozzo and Silvia Centanni

Summary. In recent years, marked point processes have found a natural application in the modeling of ultra-high-frequency financial data. The use of these processes does not require the integration of the data which is usually needed by other modeling approaches. Two main classes of marked point processes have so far been proposed to model financial data: the class of the autoregressive conditional duration models of Engle and Russel and the broad class of doubly stochastic Poisson processes with marks. In this paper, we show how to model an ultra-high-frequency data set relative to the prices of the future on the S&P 500 index using a particular class of doubly stochastic Poisson process. Our models allow a natural interpretation of the underlying intensity in terms of news reaching the market and does not require the use of ad hoc methods to treat the seasonalities. Filtering and estimation are carried out by Monte Carlo expectation maximization and reversible jump Markov chain Monte Carlo algorithms.

Key words: Cox process; Marked point process; Monte Carlo expectation maximization; Reversible jump Markov chain Monte Carlo; Shot noise process; Tick by tick data.

1 Introduction

Recently, marked point processes (MPP) have found application in the modeling of ultra-high-frequency (UHF) financial data. In UHF databases, for each market event, such as a trade or a quote update by a market maker, the time at which it took place, together with its characteristics, for instance, the price and volume of the transaction, is recorded [GDD99]. Two main classes of models based on MPP have so far been proposed for these data: the class of autoregressive conditional duration (ACD) models [ER98] and the class based on doubly stochastic Poisson processes (DSPP) with marks [RS00].

In this paper, we consider the modeling of an UHF data set relative to the prices of the future on the Standard & Poor's 500 (S&P 500) stock price index (SPU01) using a class of marked DSPP in which the stochastic part of the intensity process has a shot noise form [CI80]. In particular, we consider an intensity process, which is a given function of a non-explosive MPP, with positive jumps that we characterize through the distribution of jump times and sizes [CM06a and CM06b]. Such an intensity

Cira Perna, Marilena Sibillo (Eds.): Mathematical and Statistical Methods in Insurance and Finance

process naturally calls for an interpretation in terms of the effect of economic news on the market.

The problem of measuring the impact of economic news on asset prices and volatility has been widely studied in the literature. For example, Pearce and Roley [PR85] using a daily data set of the S&P 500 index, find that stock prices react to unexpected macroeconomic news. On the other hand, Andersen and Bollerslev [AB98] use data obtained by sampling every 5 minutes to investigate the effect of economic news on foreign exchange rates. These last authors clearly show that macroeconomic news can be linked to periods of high volatility, although they find that these news have little lasting effects. Very recently, Peters [P06] models the 2 minute frequency prices of the future on the S&P 500 index with a diffusion with piecewise constant volatility and gives a procedure for estimating the times at which the volatility changes. Even in this work, a link between economic news and prices is exploited. These and other works, however, avoid the problem of modeling UHF data, so loosing the information incorporated in the times of market events.

The paper is organized as follows. In Section 2 we introduce our modeling framework, and in Section 3 we analyze an UHF data set on the prices of the future on the S&P 500 index.

2 DSPP with Generalized Shot Noise Intensity

Let us denote with T_i and Z_i, $i = 1, 2, \ldots$, the times and sizes of the ith log return change, that is, $Z_i = \ln(S_{T_i}/S_{T_{i-1}})$, where S_t is the price in t of the future on the S&P 500 index. Then, for a filtered probability space $(\Omega, \mathcal{F}, \mathrm{P}, \{\mathcal{F}_t\}_{t\in[0,T]})$, we assume $(T_i, Z_i)_{i=0,1,2,\ldots}$, with $T_0 = 0$ and $Z_0 = 0$, be an MPP, and in particular a marked DSPP with intensity process δ (Fig. 1 shows a simulated partial realization of an MPP $(T_i, Z_i)_{i=0,1,2,\ldots}$). We will assume that $\delta_t = a(t) + b\lambda_t$, where $a(\cdot)$ is an integrable $\mathbb{R}^+$-valued deterministic function, b is a non-negative parameter, and the process λ

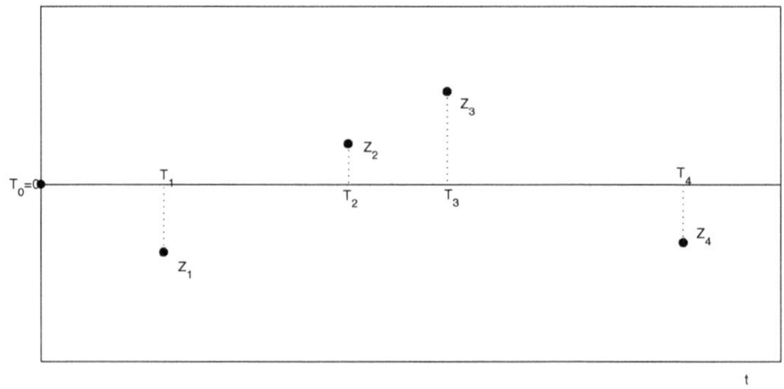

Fig. 1. The beginning of a typical trajectory of a marked point process $(T_i, Z_i)_{i=0,1,2\ldots}$, with $T_0 = 0$ and $Z_0 = 0$. In our context, T_i and Z_i represent the times and sizes of the ith log return change of the future on the S&P 500 index.

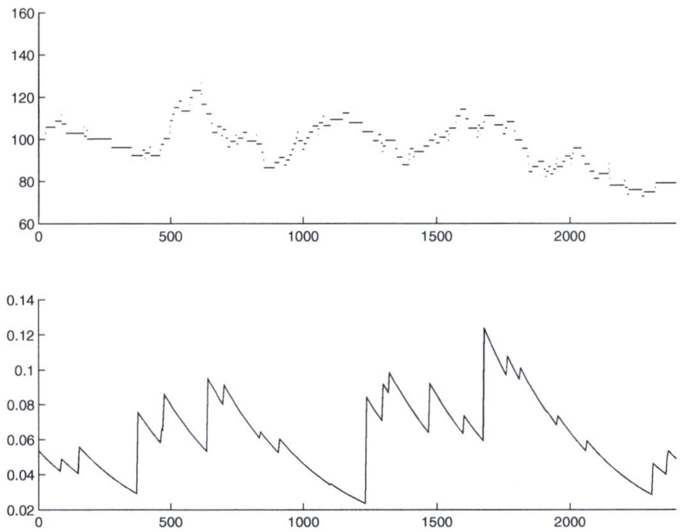

Fig. 2. Simulated trajectory (top) of the price process $S_t = S_0 \exp(\sum_{i=1}^{N_t} Z_i)$, with starting value $S_0 = 100$, time horizon $T = 2400$ and where $N_t = \#\{i \in \mathbf{N} : T_i \le t\}$ (here the price is assumed constant until the next event), obtained by conditioning to a simulated trajectory of the intensity process λ (bottom), assuming a model in the basic class with $\nu = 1/150$, $k = 0.003$ and $\gamma = 45.89$, and $a(t) = 0$ and $b = 1$.

has the form

$$\lambda_t = \sum_{j=0}^{N'_t} X_j \, e^{-k(t-\tau_j)}, \tag{1}$$

where $(\tau_j, X_j)_{j=0,1,2,\ldots}$, with $\tau_0 = 0$, is another MPP with $X_j > 0$, and where $N'_t = \#\{j \in \mathbf{N} : \tau_j \le t\}$. For this latter MPP, we also assume that the conditional distributions $p(\tau_j | \tau_1, \ldots, \tau_{j-1})$ and $p(X_j | \tau_1, \ldots, \tau_j, X_1, \ldots, X_{j-1})$ admit density.

A particularly simple specification belonging to our modeling framework is given by the basic class described by Centanni and Minozzo [CM06 and CM06b] in which the log return Z_i are assumed to be independently distributed and the parameters of the process λ are the mean inter-arrival time ν and the mean size of market perturbations $1/\gamma$. Fig. 2 shows a simulated realization of the intensity and the price process under a model in the basic class where $a(t) = 0$ and $b = 1$.

A more complicated model, always belonging to our modeling framework, is the following variation of a model by Barndorff-Nielsen and Shephard [BS01] in which the intensity process is given by λ in (1), and where

i) τ_1 is exponentially distributed with mean δ_1; $(\tau_j - \tau_{j-1})$, $j = 2, 3, \ldots$, are conditionally distributed, given X_{j-1}, as exponentials with means δ / X_{j-1};
ii) $(X_0 - 1)$ is Poisson distributed with mean η_0; $(X_j - 1)$, $j = 1, 2, \ldots$, are Poisson distributed with mean η;

iii) $(T_i)_{i\in\mathbb{N}}$ are the event times of a marked DSPP with intensity λ;

and

iv) $(Z_i)_{i\in\mathbb{N}}$ are conditionally independently distributed as Gaussians with mean $\mu(T_i - T_{i-1}) + \alpha \int_{T_{i-1}}^{T_i} \lambda_t \,\mathrm{d}t$, and variance σ^2.

With respect to other modeling approaches, our modeling framework has some advantages as well as some other good features. First of all, data do not need to be sampled at fixed frequency, indeed, the inter-arrival time between price changes gives us important information that is accounted for in the model. Secondly, the form assumed by the intensity process allows a natural interpretation of the stochastic changes of the process λ in terms of market perturbations caused by the arrival of relevant news [KLP04]. When the jth item of news reaches the market, a sudden increase X_j in trading activity occurs, depending on the importance of the item, followed by a progressive normalization. The random variable τ_j represents the time of the arrival of the jth item of news, whereas the parameter k expresses the speed of absorption of the effect of the news by the market. Another good feature of our modeling approach is the possibility to account for intra-day seasonal patterns through the specification of the model, without the need to resort to 'ad hoc' methods. Indeed, the function $a(\cdot)$ can be interpreted as the activity that the market would have in absence of random perturbations. By adequately choosing $a(\cdot)$, it is possible to model many of the stylized features characterizing intra-day price data [GDD99].

As far as inference is concerned, for models belonging to our class, likelihood inference on the parameters can be performed by Monte Carlo expectation maximization (MCEM) algorithms [FM03]. Assuming that the distribution of Z_i, τ_j, X_j depends on a parameter vector θ, and by identifying trajectories λ_0^T with vectors $(N'_T, \tau, \mathbf{X}) = (N'_T, \tau_0, \ldots, \tau_{N'_T}, X_0, \ldots, X_{N'_T})$, the full likelihood can be written as

$$f(\mathbf{Z}, \mathbf{T}, N, \mathbf{X}, \tau, N'; \theta)$$
$$= f(\mathbf{Z}|\mathbf{T}, N, \mathbf{X}, \tau, N'; \theta)\ \mathrm{P}(T_{N+1} > T|\mathbf{T}, \mathbf{X}, \tau, N'; \theta)\ f(\mathbf{T}|\mathbf{X}, \tau, N'; \theta)\ f(\mathbf{X}, \tau, N'; \theta).$$

Then, if the guess for θ at the rth iteration ($r = 1, 2, \ldots, R$) is θ_r, the MCEM algorithm proceeds as follows:

S step draw S_r samples $(\mathbf{X}, \tau, N')^{(s)}$, $s = 1, \ldots, S_r$, from the conditional distribution $p(\mathbf{X}, \tau, N'|\mathbf{Z}, \mathbf{T}, N; \theta_r)$;

E step $$Q_r(\theta, \theta_r) = \frac{1}{S_r}\sum_{s=1}^{S_r} \ln f(\mathbf{Z}, \mathbf{T}, N, \mathbf{X}^{(s)}, \tau^{(s)}, N'^{(s)}; \theta);$$

M step take as the new guess θ_{r+1} the value of θ which maximizes $Q_r(\theta, \theta_r)$.

By specifying specific values for S_r, $r = 1, 2, \ldots, R$, we get particular versions of the algorithm. For $S_r = 1$ we have the stochastic EM (StEM) algorithm. This algorithm does not converge pointwise and estimates are given by some summary statistic. For S_r large, the algorithm behaves similarly to the (deterministic) EM algorithm, whereas for S_r increasing with r, we have a simulated annealing MCEM.

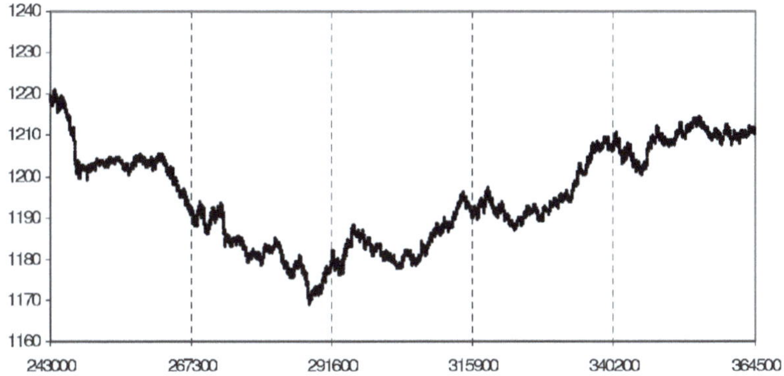

Fig. 3. Prices of the future on the S&P 500 index from the 23rd to the 27th of July 2001 (the last five trading days out of the 15 considered in the analysis). Overall, there were 39,889 price changes in the three weeks considered (time is expressed in seconds).

For many marked DSPP with generalized shot noise intensity, the almost sure convergence of the (simulated annealing) MCEM can be derived from the results of Fort and Moulines [FM03] for curved exponential families.

The implementation of this algorithm crucially relies on the existence of an efficient sampler for the stochastic step in which we need to sample from the conditional distribution $p(\mathbf{X}, \tau, N'|\mathbf{Z}, \mathbf{T}, N; \theta_r)$. Let us note that such a sampler provides a Monte Carlo solution to the filtering (smoothing) problem. Details on how to build a sampler based on the reversible jump Markov chain Monte Carlo (RJMCMC) are given in [CM06b]. A sampler similar in spirit to ours has been studied by [RPD04].

3 The S&P 500 Index Future Data Set

We considered all price changes (39,889) from the 9th to the 27th of July 2001 (15 trading days). In Fig. 3 we report the last five days of the period considered (time is expressed in seconds). The market was open from 8.30 in the morning to 15.15 in the afternoon. The data set showed successive log returns not to be autocorrelated and also not correlated with the waiting times $T_i - T_{i-1}$. Moreover, the time between successive price changes showed to be exponentially distributed. To account for the intra-day seasonality in the data, every one of the 15 days considered (each day corresponds to 405 minutes, for a total time horizon of $T = 6{,}075$ minutes) has been sliced into 27 subintervals (15 minutes long). The number of changes of the price of the S&P 500 index future within each of the 15 minutes intervals has been calculated. The circles in Fig. 4 show, for the 27 intra-day intervals, the minimum number of changes of the price of the future (in the corresponding interval) over the 15 days. The solid line, obtained from the least square fit (of a second order polynomial) by subtracting a constant equal to 1.2, can be assumed to provide the deterministic part of the intensity process accounting for the intra-day seasonality. Thus, to model the

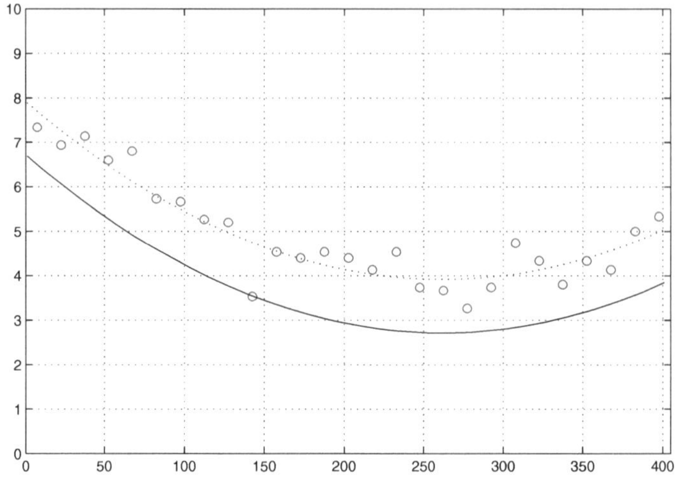

Fig. 4. Detrending of the intra-day seasonality in the S&P 500 future data set. Every one of the 15 days considered (each day corresponding to 405 minutes, for a total of $T = 6{,}075$ minutes) has been sliced into 27 intervals (15 minutes long). The number of changes of the price of the S&P 500 index future within each of the 15 minute intervals has been calculated. The circles in the graph report, for the 27 intra-day intervals, the minimum number of changes of the price (in the corresponding interval) over the 15 days. The dotted line is the least square fit of a second order polynomial. The solid line is obtained from the least square fit by subtracting a constant value equal to 1.2, and is used as a deterministic 'offset' for the intensity process.

data we consider an intensity of the form, for $t \geq 0$, $a_0 + a_1 t + a_2 t^2 + \lambda_t$, where $a_0 = 6.7226$, $a_1 = -0.0306$ and $a_2 = 0.0001$. Let us note that the chosen shape for the deterministic part of the intensity is standard in UHF data analysis (see, for instance, Fig. 4 of Rydberg and Shephard [RS00]).

As usual in financial applications where it is reasonable to assume a partial information setting in which market agents are restricted to observe only the history of the stock price, we assume that past times and sizes of price changes of the S&P 500 index future have been observed, but not the intensity process. Under the assumptions of the basic class, estimates of the parameters $\theta = (k, \gamma, \nu)$ were obtained by implementing a StEM algorithm as in Centanni and Minozzo [CM06b]. Fig. 5 shows a run of 450 steps of the StEM algorithm (each step involving 6,000 updates of an RJMCMC). Only the 6,000th sampled intensity is used in E and M steps of the StEM. This is also used as the initial intensity in the RJMCMC at the subsequent step of the StEM. Initial values for the StEM algorithm are $k = 0.001$, $\gamma = 25$ and $\nu = 0.1$. Point estimates of k, γ and ν are provided by the sample means over the last 150 steps of the StEM algorithm, and give $\hat{k} = 0.006$, $\hat{\gamma} = 2.875$ and $\hat{\nu} = 0.050$.

The filtering (smoothing) of the intensity was performed by running an RJMCMC algorithm assuming the above estimates as the true parameter values. The top graph in Fig. 6 shows the smoothing expectation $\mathrm{E}(\lambda_t | \mathbf{Z}, \mathbf{T}, N)$, $t = 0, 5, 10, 15, \ldots, T = 6{,}075$. After a "burn in" period of 10,000, the RJMCMC algorithm was

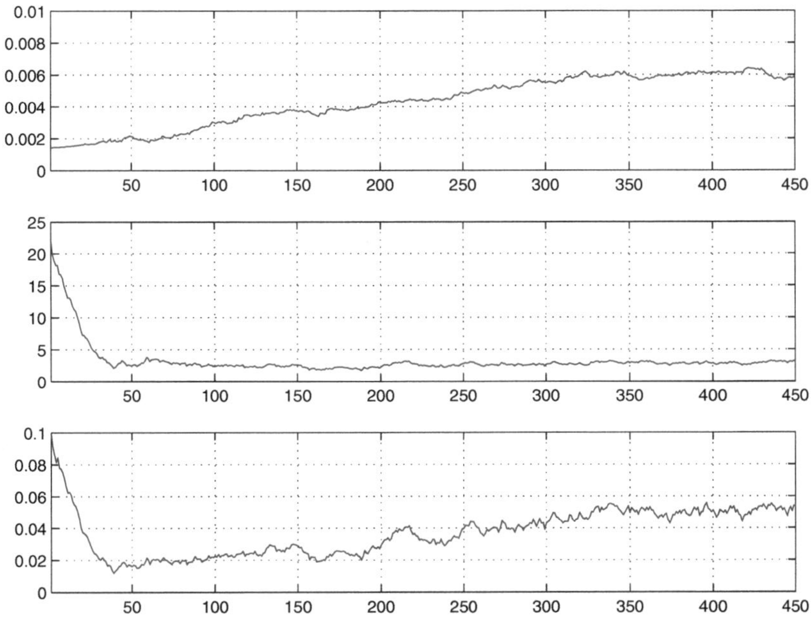

Fig. 5. Estimation of the parameter θ with the StEM algorithm for the S&P 500 future data set. Monitoring of k (top), γ (middle) and ν (bottom), assuming a model in the basic class with a deterministic seasonality for the intensity, for a run of 450 steps of the StEM algorithm (each step involving 6,000 RJMCMC updates). Initial values for the StEM algorithm are $k = 0.001$, $\gamma = 25$ and $\nu = 0.1$. Point estimates of k, γ and ν are provided by the sample means over the last 150 steps of the StEM algorithm and give $\hat{k} = 0.006$, $\hat{\gamma} = 2.875$ and $\hat{\nu} = 0.050$.

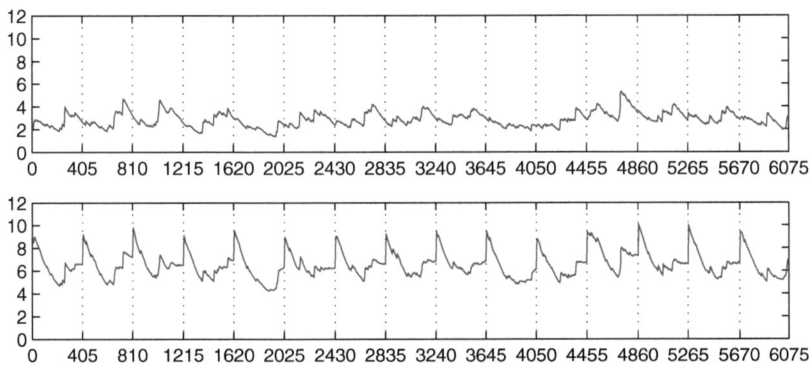

Fig. 6. The top graph shows the filtering expectation $\mathrm{E}(\lambda_t|\mathbf{Z}, \mathbf{T}, N)$, $t = 0, 5, 10, 15, \ldots, T = 6{,}075$, (time is expressed in minutes) under the model described in the text, assuming the parameter values: $k = 0.006$, $\gamma = 2.875$, $\nu = 0.050$. After a "burn in" period of 10,000, the RJMCMC algorithm was run for other 10,000 updates, recording a trajectory every 2. The bottom graph shows the total (smoothed) intensity comprehensive also of the deterministic part given in Fig. 4.

run for other 10,000 updates, recording a trajectory every 2. The bottom graph shows the total (smoothed) intensity comprehensive also of the deterministic part shown in Fig. 4. The peaks in the top graph show periods of unexpected high trading activity, not accounted for by the intra-day seasonality, which may correspond to the arrival of relevant news. The smoothed intensity can also be used in the numerical calculation of quantities of interest in a number of financial problems, such as price forecasting.

References

[AB98] Andersen, T., Bollerslev, T.: Deutsche Mark-Dollar volatility: intraday activity patterns, macroeconomic announcements, and longer run dependencies. The Journal of Finance, **53**, 219–265 (1998)

[BS01] Barndorff-Nielsen, O. E., Shephard, N.: Non-Gaussian Ornstein-Uhlenbeck-based models and some of their uses in financial economics (with discussion). Journal of the Royal Statistical Society, Series B, **63**, 167–241 (2001)

[CM06a] Centanni, S., Minozzo, M.: Estimation and filtering by reversible jump MCMC for a doubly stochastic Poisson model for ultra-high-frequency financial data. Statistical Modelling, **6**, 97–118 (2006)

[CM06b] Centanni, S., Minozzo, M.: A Monte Carlo approach to filtering for a class of marked doubly stochastic Poisson processes. Journal of the American Statistical Association, **101**, 1582–1597 (2006)

[CI80] Cox, D. R., Isham, V.: Point Processes. Chapman and Hall, London (1980)

[ER98] Engle, R. F., Russel, J.R.: Autoregressive conditional duration: A new model for irregularly spaced transaction data. Econometrica, **66**, 1127–1162 (1998)

[FM03] Fort, G., Moulines, E.: Convergence of the Monte Carlo expectation maximization for curved exponential families. The Annals of Statistics, **31**, 1220–1259 (2003)

[GDD99] Guillaume, D. M., Dacorogna, M. M., Davé, R. R., Müller, U. A., Olsen, R. B., Pictet, O. V.: From the bird's eye to the microscope: A survey of new stylized facts of the intra-daily foreign exchange markets. Finance and Stochastics, **1**, 95–129 (1999)

[KLP04] Kalev, P. S., Liu, W. M., Pham, P. K., Jarnecic, E.: Public information arrival and volatility of intraday stock returns. Journal of Banking & Finance, **28**, 1441–1467 (2004)

[P06] Peters, R.T.: Shocks in the S&P 500. Working paper, University of Amsterdam (2006)

[PR85] Pierce, D.K., Roley, V. V.: Stock prices and economics news. Journal of Business, **58**, 49–67 (1985)

[RPD04] Roberts, G.O., Papaspiliopoulos, O., Dellaportas, P.: Bayesian inference for non-Gaussian Ornstein-Uhlenbeck stochastic volatility processes. Journal of the Royal Statistical Society, Series B, **66**, 369–393 (2004)

[RS00] Rydberg, T. H., Shephard, N.: A modelling framework for the prices and times of trades made on the New York stock exchange. In: Fitzgerald, W.J., Smith, R.L., Walden, A. T., Young, P. C. (eds) Nonlinear and Nonstationary Signal Processing, pp. 217–246 (2000)

Simulating a Generalized Gaussian Noise with Shape Parameter 1/2

Martina Nardon and Paolo Pianca

Summary. This contribution deals with Monte Carlo simulation of generalized Gaussian random variables. Such a parametric family of distributions has been proposed for many applications in science and engineering to describe physical phenomena. Its use also seems interesting in modeling economic and financial data. For low values of the shape parameter α, the distribution presents heavy tails. In particular, $\alpha = 1/2$ is considered and for such a value of the shape parameter, different simulation methods are assessed.

Key words: Generalized Gaussian density; Heavy tails; Transformations of random variables; Monte Carlo simulation; Lambert W function.
JEL Classification Numbers: C15, C16.
MathSci Classification Numbers: 33B99, 65C05, 65C10.

1 Introduction

The parametric family of generalized Gaussian densities has been used to model successfully physical phenomena, and in engineering is used, for example, in the area of signal processing (see [KN05]).

Some important classes of random variables belong to this family such as, amongst others, the Gaussian distribution and the Laplacian distribution. Moreover, for values of the shape parameter within a certain interval (which is of interest in many practical situations), the moments of such distributions are all finite.

Obviously, in many applications, accurate and fast simulation of a stochastic process of interest can play an important role. In this contribution, we analyze different methods for generating deviates from a generalized Gaussian distribution.

In the special case of the shape parameter $\alpha = 1/2$, we compare the efficiency of four different simulation techniques. Numerical results highlight that the technique based on the inverse cumulative distribution function written in terms of the Lambert W function is the most efficient.

An outline of the paper is as follows. Section 2 presents some properties of the generalized Gaussian density. In sections 3 and 4 simulation techniques for general-

Cira Perna, Marilena Sibillo (Eds.): Mathematical and Statistical Methods in Insurance and Finance

ized Gaussian random variables are analyzed. In particular, $\alpha = 1/2$ is considered. Section 5 concludes.

2 The Generalized Gaussian Density

The probability density function of a generalized Gaussian random variable X, with mean μ and variance σ^2, is defined as

$$f_X(x;\ \mu, \sigma, \alpha) = \frac{\alpha}{2} \frac{A(\alpha, \sigma)}{\Gamma(1/\alpha)} \exp\left\{-\left(A(\alpha, \sigma)\,|\,x - \mu\,|\right)^{\alpha}\right\} \qquad x \in \mathbb{R}, \tag{1}$$

where

$$A(\alpha, \sigma) = \frac{1}{\sigma}\left[\frac{\Gamma(3/\alpha)}{\Gamma(1/\alpha)}\right]^{1/2} \tag{2}$$

and $\Gamma(z) = \int_0^{+\infty} t^{z-1} e^{-t} dt$ $(z > 0)$ is the complete gamma function.

The generalized Gaussian distribution (GGD) is symmetric with respect to μ. $A(\alpha, \sigma)$ is a scaling factor which defines the dispersion of the distribution, hence it is a generalized measure of the variance. $\alpha > 0$ is the shape parameter which describes the exponential rate of decay: heavier tails correspond to smaller values of α.

The generalized Gaussian family includes a variety of random variables. Some well known classes of distributions are generated by a parametrization of the exponential decay of the GGD. When $\alpha = 1$, the GGD corresponds to a Laplacian, or double exponential distribution. For $\alpha = 2$ one has a Gaussian distribution. When $\alpha \to +\infty$ the GGD converges to a uniform distribution in $(\mu - \sqrt{3}\sigma,\ \mu + \sqrt{3}\sigma)$, while when $\alpha \to 0^+$ we have an impulse probability function at $x = \mu$.

All the odd central moments of distribution (1) are zero, $\mathbb{E}(X - \mu)^r = 0$ $(r = 1, 3, 5, \ldots)$, and the even central moments are

$$\mathbb{E}(X - \mu)^r = \left[\frac{\sigma^2 \Gamma(1/\alpha)}{\Gamma(3/\alpha)}\right]^{r/2} \frac{\Gamma((r+1)/\alpha)}{\Gamma(1/\alpha)} \qquad r = 2, 4, 6, \ldots\,. \tag{3}$$

With a straightforward standardization and some reductions from (1), we obtain the following GGD with zero-mean and unit-variance

$$f_X(x;\ \alpha) = \frac{\alpha}{2} \frac{A(\alpha)}{\Gamma(1/\alpha)} \exp\left\{-\left(A(\alpha)\,|\,x\,|\right)^{\alpha}\right\}, \tag{4}$$

where $A(\alpha) = A(\alpha, 1)$.

In the following, we confine our attention to generalized Gaussian random variables with density (4). For $0 < \alpha < 2$ the density (4) is suitable for modeling many physical and financial processes with heavy tails. It is worth noting that, for GGD with $0 < \alpha < 2$, all the moments are finite (this is not the case for other heavy-tailed densities, e.g. stable densities).

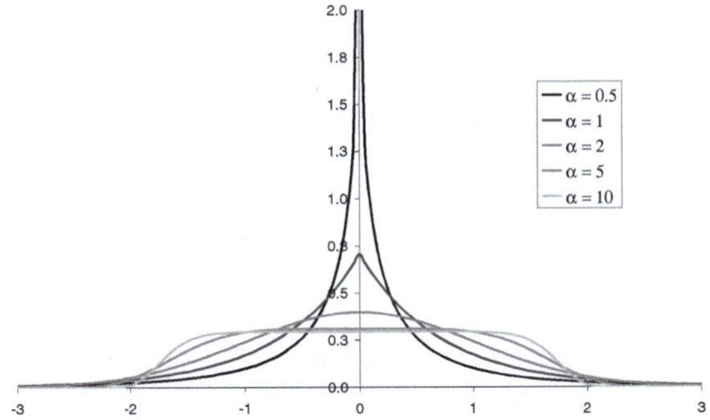

Fig. 1. Generalized Gaussian densities for different values of the parameter α, with $\mu = 0$ and $\sigma = 1$.

The kurtosis of density (4) is

$$\mathcal{K}(\alpha) = \frac{\Gamma(1/\alpha)\,\Gamma(5/\alpha)}{\left[\Gamma(3/\alpha)\right]^2}\,. \tag{5}$$

$\mathcal{K}(\alpha)$ decreases with α; moreover, the following results hold: $\lim_{\alpha\to 0^+} \mathcal{K}(\alpha) = +\infty$ and $\lim_{\alpha\to+\infty} \mathcal{K}(\alpha) = 1.8$ (see [DMGRD01]). Fig. 1 shows the generalized Gaussian densities for different values of the parameter α, with zero mean and unit variance.

3 Simulating the Generalized Gaussian Distribution

Let $F_X(\cdot)$ be the cumulative distribution function (cdf) of a random variable X and $F_X^{-1}(\cdot)$ be its inverse. It is well known that if $F_X^{-1}(\cdot)$ can be directly evaluated, a large number of realizations of X can be obtained as $x_i = F_X^{-1}(u_i)$, where u_i ($i = 1, 2, \ldots, n$) are random numbers uniform over (0, 1). If F_X^{-1} has a closed form expression, such a method can be applied efficiently. Unfortunately, this is not always the case. Nevertheless, if $F_X(\cdot)$ can be evaluated, it can still be possible to simulate the random variable X by numerically inverting its cumulative distribution function. When $F_X(\cdot)$ has no closed form expression, numerical integration or other approximation methods are required, at the expense of an increasing computational amount. An alternative simulation method is based on techniques of transformation of a random variable X for which a random number generator is available. A specific technique will be described below for the GGD case.

Let X be a generalized Gaussian random variable with cumulative distribution function

$$F_X(x) = \int_{-\infty}^{x} \frac{\alpha}{2}\frac{A(\alpha)}{\Gamma(1/\alpha)} \exp\left\{-\left(A(\alpha)\,|\,t\,|\right)^{\alpha}\right\} dt\,, \tag{6}$$

where $A(\alpha)$ has been defined above. Such a function can be written in closed form only in a very few special cases.

In order to generate values from a generalized Gaussian distribution with parameter α, the following three-step procedure can be used:

i) Simulate a gamma random variable $Z \sim Gamma(a, b)$, with parameters $a = \alpha^{-1}$ and $b = (A(\alpha))^{\alpha}$.
ii) First apply the transformation $Y = Z^{1/\alpha}$.
iii) Finally, apply a transformation of the form

$$Y = |\, X \,| \,. \tag{7}$$

Relationship (7) has two roots. The main problem is how to determine the probability for choosing one of such roots. It can be shown (see [MSH76]) that, as a consequence of the symmetry of the GG distribution, one takes the roots with equal probability. For each random observation y, a root is chosen ($x = -y$ or $x = y$). To this end, an auxiliary Bernoulli trial with probability $p = 1/2$ is performed.

We first encountered the problem of generating random variates from the gamma distribution[1]. The process relies on the assumption that if Z is a $Gamma(a, b)$ distributed random variable (with a and b as defined above), then by letting $Y = Z^{1/\alpha}$ and considering the transformation (7), as a result, X has a GGD distribution with parameter α.

In testing this three-step procedure, a number of large samples were generated with various choices for the parameter α.

4 Simulating the Generalized Gaussian Distribution with $\alpha = 1/2$

With regard to density (4), if we address the case $\alpha = 1/2$ we obtain the GG density

$$f_X(x) = \frac{\sqrt{30}}{2} \exp\left\{-2\sqrt{30}\,|\, x\,|^{1/2}\right\} \tag{8}$$

and the cumulative distribution function

$$F_X(x) = \begin{cases} \frac{1}{2}\exp\left\{-2\sqrt{30}\,(-x)^{1/2}\right\}\left(1 + 2\sqrt{30}\,(-x)^{1/2}\right) & x \le 0 \\ 1 - \exp\left\{-2\sqrt{30}\,x^{1/2}\right\}\frac{1}{2}\left(1 + 2\sqrt{30}\,x^{1/2}\right) & x > 0\,. \end{cases} \tag{9}$$

In order to generate variates from a GG distribution, we can apply the method based on the transformation described in the previous section. Otherwise, realizations of a random variable with cdf (9) can be obtained as $x = F_X^{-1}(u)$, where u are random numbers from a uniform distribution on $(0, 1)$. The function $F_X(\cdot)$ can be inverted using numerical techniques.

The inverse function of the cdf (9) can be expressed in terms of the so-called *Lambert W function*. Based on such a result, a very fast and accurate simulation procedure can be defined. The following subsection is devoted to the definition of the Lambert W function and the description of its main properties.

[1] In the numerical experiments we used the routine DRNGAM of the IMSL library for the generation of deviates from a distribution $Gamma(a, 1)$.

4.1 The Lambert W function

The Lambert W function, also known as the *Omega* function, is implicitly specified as the root of the equation

$$W(z)e^{W(z)} = z\,. \tag{10}$$

It is a multivalued function defined in general for z complex and assuming values $W(z)$ complex. If z is real and $z < -1/e$, then $W(z)$ is multivalued complex. If $z \in \mathbb{R}$ and $-1/e \le z < 0$, there are two possible real values of $W(z)$: the branch satisfying $W(z) \ge -1$ is usually denoted by $W_0(z)$ and called the *principal branch* of the W function, and the other branch satisfying $W(z) \le -1$ is denoted by $W_{-1}(z)$. If $z \in \mathbb{R}$ and $z \ge 0$, there is a single value for $W(z)$ which also belongs to the principle branch $W_0(z)$. The choice of solution branch usually depends on physical arguments or boundary conditions[2].

A high-precision evaluation of the Lambert W function is available in *Maple* and *Mathematica* softwares. In particular, *Maple* computes the real values of W using the third-order Halley's method (see [A81]), giving rise to the following recursion

$$w_{j+1} = w_j - \frac{w_j\, e^{w_j} - z}{(w_j+1)\, e^{w_j} - \frac{(w_j+2)(w_j\, e^{w_j} - z)}{2w_j+2}} \qquad j \ge 0\,. \tag{11}$$

Analytical approximations of the W function are also available and can be used as an initial guess in the iterative scheme (11) (see [CGHJK96] and [CBM02]).

In the numerical experiments carried out, we applied Halley's method and as an initial guess we adopted

$$w_0 = \begin{cases} -1+\rho-\dfrac{1}{3}\rho^2+\dfrac{11}{72}\rho^3-\dfrac{43}{540}\rho^4+\dfrac{769}{17\,280}\rho^5-\dfrac{221}{8\,505}\rho^6 & -\dfrac{1}{e} \le z < -0.333 \\[2ex] \dfrac{-8.096+391.0025z-47.4252z^2-4877.633z^3-5532.776z^4}{1-82.9423z+433.8688z^2+1515.306z^3} & -0.333 \le z \le -0.033 \\[2ex] L_1-L_2+\dfrac{L_2}{L_1}+\dfrac{L_2(-2+L_2)}{2L_1^2}+\dfrac{L_2(6-9L_2+2L_2^2)}{6L_1^3}+ & \\ +\dfrac{L_2(-12+36L-2-22L_2^2+3L_2^3)}{12L_1^4}+ & -0.033 < z < 0 \\ +\dfrac{L_2(60-300L_2+350L_2^2-125L_2^3+12L_2^4)}{60L_1^5} & \end{cases} \tag{12}$$

where $\rho = -\sqrt{2(e\,z+1)}$, $L_1 = \ln(-z)$, and $L_2 = \ln[\ln(-z)]$. Formula (12) used as a direct method to approximate the function $W_{-1}(z)$ for $z \in [-1/e,\ 0)$ provides quite accurate values. Indeed, the maximum relative error is below 10^{-4} (see [CBM02]).

4.2 Analytical inversion of GGD with $\alpha = 1/2$

The aim here is to invert the cumulative distribution function (9) and prove that this inverse can be expressed in terms of the Lambert W function. We first consider the

[2] The paper [CGHJK96] reports the main properties and applications of such a function.

cumulative distribution function for $x \leq 0$. Setting $y = -(-2\sqrt{30}x)^{1/2}$, we have

$$F = \frac{1}{2}(1-y)e^y \quad (13)$$

i.e.

$$-\frac{2F}{e} = (1-y)e^{y-1}. \quad (14)$$

Equation (14) leads to

$$y - 1 = W_{-1}(-2F/e) \quad (15)$$

and therefore

$$x = -\frac{1}{2\sqrt{30}}[1 + W_{-1}(-2F/e)]^2. \quad (16)$$

The correct branch W_{-1} is identified from the boundary conditions ($x = 0$, $F = 1/2$) and ($x \to -\infty$, $F \to 1/2$).

Similar arguments permit consideration of the case $x \geq 0$. Thus, we have the inverse cumulative distribution function

$$F^{-1}(u) = \begin{cases} -\frac{1}{2\sqrt{30}}\left[1 + W_{-1}(-2u/e)\right]^2 & 0 < u \leq 1/2 \\ -\frac{1}{2\sqrt{30}}\left[1 + W_{-1}(-2(1-u)/e)\right]^2 & 1/2 < u < 1\,. \end{cases} \quad (17)$$

Equation (17) solves the problem of obtaining a large number of realizations of generalized Gaussian density with $\alpha = 1/2$, provided we are able to compute the function W_{-1}. Based on such results, a fast and fairly accurate simulation procedure can be defined.

4.3 Numerical results

Realizations of a generalized Gaussian random variable with $\alpha = 1/2$, generated with the algorithm based on the inverse cumulative distribution function (17) and the approximation of the Lambert W are shown in Fig. 2.

We have performed a Monte Carlo estimation of probability density (9) based on 10^8 values, collected into bins of width $\Delta x = 0.05$. Fig. 3 shows the simulated

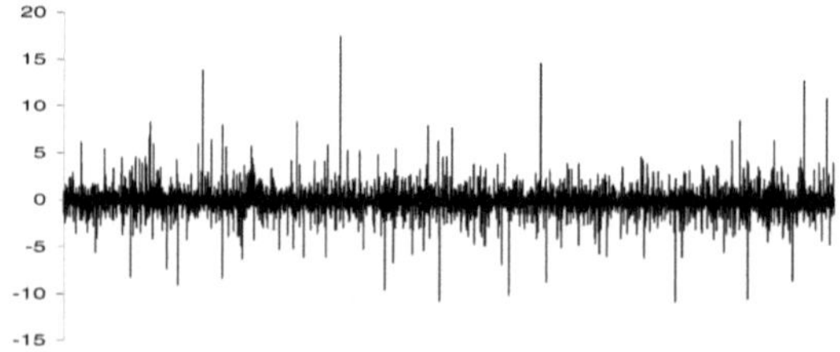

Fig. 2. Instances of GGD with parameter $\alpha = 1/2$, $\mu = 0$ and $\sigma = 1$ ($N = 10000$).

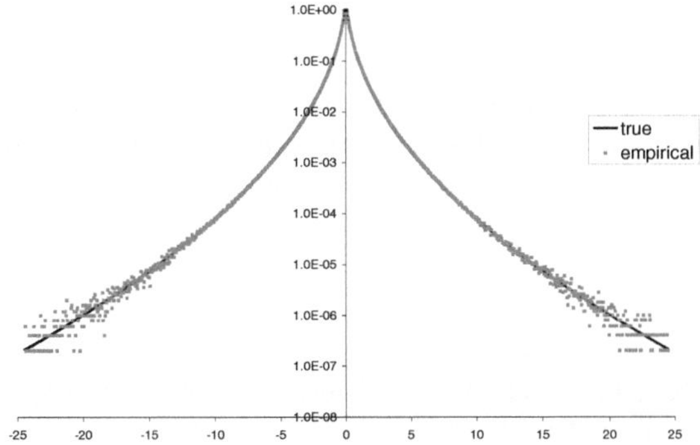

Fig. 3. Estimated and theoretical probability density functions of a generalized Gaussian random variable with $\alpha = 1/2$.

probability density function and the theoretical density (8). As can be seen in the figure, the probability density function is well approximated by the empirical density. The Kolmogorov-Smirnov test yielded no evidence indicating that the simulated observations were not generated from the GGD distribution.

Table 1 reports the estimated moments, based on the 10^6 simulations, of the GGD and the computational time. The method based on the inverse cdf written in terms of the Lambert W function and approximation (12) is very fast. The results of the simulation based on the inverse cdf are not reported. Such a simulation method is very inefficient with respect to the other techniques.

5 Conclusions

In this contribution, we investigated and compared different techniques for generating variates from a generalized Gaussian distribution. For $\alpha = 1/2$ a fast and accu-

Table 1. Simulating the generalized Gaussian distribution with $\alpha = 1/2$ $(N = 10^6)$.

Simulation method	Mean	Variance	Kurtosis $\mathcal{K} = 25.2$	cpu time (sec.)
Lambert W & approximation (12)	−0.000101	0.999951	25.396091	1.97
Lambert W & Halley's algorithm	−0.000101	0.999969	25.395066	4.10
Three-step procedure	−0.000077	1.000630	25.103438	10.77

rate simulation procedure can be defined. A three-step procedure based on a gamma transformation can also be successfully applied.

Generalized Gaussian distributions have been proposed in many applications in science and in engineering. It seems interesting to investigate their possible applications in modeling economic and financial data. This is left for future research.

References

[A81] Alefeld, G.: On the convergence of Halley's method. The American Mathematical Monthly, **88**(7), 530–536 (1981)

[CBM02] Chapeau-Blondeau, F., Monir, A.: Numerical evaluation of the Lambert W function and application to generation of generalized Gaussian noise with exponent 1/2. IEEE Transactions on Signal Processing, **50**(9), 2160–2165 (2002)

[CGHJK96] Corless, R. M., Gonnet, G. H., Hare, D. E. G., Jeffrey, D. J., Knuth, D. E.: On the Lambert W function. Advances in Computational Mathematics, **5**, 329–359 (1996)

[DMGRD01] Domínguez-Molina, J. A., González-Farías, G., Rodríguez-Dagnino, R. M.: A practical procedure to estimate the shape parameter in the generalized Gaussian distribution. Technique Report, I-01-18 (`http://www.cimat.mx`) (2001)

[KN05] Kokkinakis, K., Nandi, A. K.: Exponent parameter estimation for a generalized Gaussian probability density functions with application to speech modeling. Signal Processing, **85**, 1852–1858 (2005)

[MSH76] Michael, J. R., Schucany, W. R., Haas, R. W.: Generating random variates using transformations with multiple roots. The American Statistician, **30**, 88–90 (1976)

Further Remarks on Risk Profiles for Life Insurance Participating Policies

Albina Orlando and Massimiliano Politano

Summary. This paper deals with the calculation of suitable risk indicators for Life Insurance policies in a Fair Value context. In particular, after determining the quantile reserve for Life Insurance Participating Policies, the role of the term structure of mortality rates is analyzed in risk determinations. Numerical results are investigated in order to determine not only suitable risk indicators, but also the mortality risk impact in this context.

Key words: Participating Policies; Fair Pricing; Quantile Reserve; Mortality Risk.

1 Introduction

At the begin of 2004, the International Accounting Standard Boards (IASB) issued the International Financial Reporting Standard 4 Insurance Contracts (IFRS 4), providing, for the first time, guidance on accounting and marking the first step in the IASB's project to achieve the convergence of widely varying insurance accounting practices around the world. In particular, IFRS 4 permits the introduction of an accounting policy consistently reflecting current interest rates and, if the insurer so elects, other current estimates and assumptions, giving rise to a potential reclassification of some or all financial assets "at fair value" when an insurer changes accounting policies for insurance liabilities. The IASB defines the Fair Value as "an estimate of the price an entity would have realized if it had sold an asset, or paid if it had been relieved a liability on the reporting date in an arm's length exchange motivated by normal business considerations": therefore, the problem is posed of determining the market value of insurance liabilities. In particular, the IASB allows for the use of stochastic models in order to estimate future cash flow. In the actuarial perspective, the introduction of an accounting policy and a fair value system implies that the mathematical reserve could be defined as the net present value toward the policyholder evaluated at current interest rates and, eventually, at current mortality rates. The paper is organized as follows: Section 2 gives a survey regarding the application of the quantile reserve to actuarial liabilities. Section 3 develops the mathematical framework for the fair valuation of the participating policy. Finally, in Section 4, numerical evidence is offered.

Cira Perna, Marilena Sibillo (Eds.): Mathematical and Statistical Methods in Insurance and Finance

2 The Quantile Reserve and the Actuarial Liabilities

The quantification of a liability fair value can be approached by introducing the replicating portfolio, that is, a portfolio of financial instruments giving origin to a cash flow matching the one underlying the liability itself. The fair valuation of insurance liabilities, since considering cash flow depending on the human life, and so not trading in an existing market, can be considered existent in economic reality, as stated in Buhlmann, and can be measured by means of a financial instrument portfolio. Within this scenario, it is possible to introduce quantitative tools such as the quantile reserve. Indicating by $W(t)$ the financial position at time t, that is the stochastic mathematical reserve of a life insurance contract, the quantile reserve at confidence level α, $0 < \alpha < 1$, is expressed by the value $W^*(t)$ in the following equation

$$P\left\{W(t) > W^*(t)\right\} = \alpha \,.$$

As one can see, the quantile reserve is a threshold value in the sense that in the $(1-\alpha)100\%$, $R(t)$ is smaller or equal to the quantile reserve.

3 The Mathematical Model

Let us consider an endowment policy issued at time 0 and maturing at time ξ , with initial sum insured C_0. Moreover, let us define $\{r_t; t = 1, \ldots, \xi\}$ and $\{\mu_{x+t}; t = 1, \ldots, \xi\}$ the random spot rate process and the mortality process respectively, both of them measurable with respect to the filtrations $\mathcal{F}^r$ and $\mathcal{F}^\mu$. The above mentioned processes are defined on a unique probability space $(\Omega, \mathcal{F}^{r,\mu}, P)$ such that $\mathcal{F}^{r,\mu} = \mathcal{F}^r \cup \mathcal{F}^\mu$. For the revaluable endowment policy, we assume that, in the case of single premium, at the end of the t-th year, if the contract is still in force, the mathematical reserve is adjusted at a rate ρ_t defined as follows

$$\rho_t = \max\left\{\frac{\eta S_t - i}{1+i}, 0\right\} \qquad t = 1, \ldots, \xi \,. \tag{1}$$

The parameter η, $0 \le \eta \le 1$, denotes the constant participating level, and S_t indicates the annual return of the reference portfolio. Equation (1) explains the fact that the total interest rate credited to the mathematical reserve during the t-th year, is the maximum between ηS_t and i, where i is the minimum rate guaranteed to the policyholder. Since we are dealing with a single premium contract, the bonus credited to the mathematical reserve implies a proportional adjustment at the rate ρ_t also of the sum insured. It is assumed that if the insured dies within the term of the contract, the benefit increases by an additional last adjustment at the end of the year of death.

Denoting by C_t, $t = 1, \ldots, \xi$, the benefit paid at time t if the insured dies between ages $x+t-1, x+t$ or, in the case of survival, for $t = \xi$ the following recursive equation holds for benefits of successive years

$$C_t = C_{t-1}(1+\rho_t) \qquad t = 1, \ldots, \xi \,.$$

The iterative expression for them is instead

$$C_t = C_0 \prod_{j=1}^{t}(1+\rho_j) \qquad t = 1, \ldots, \xi\,,$$

where we have indicated by ϕ_t the re-adjustment factor

$$\phi_t = \prod_{j=1}^{t}(1+\rho_j) \qquad t = 1, \ldots, \xi\,.$$

In this context, as the elimination of the policyholder can happen in the case of death in the year $t \in [0, \xi[$. or in the case of survival $t = \xi$, the liability borne out by the insurance company can be expressed in this manner

$$W_0^L = \sum_{t=0}^{\xi} C_{t\,t-1/1}Y_x + C_{\xi\,\xi}J_x\,, \tag{2}$$

where ${}_{t-1/1}Y_x = \begin{cases} e^{-\Delta(t)} & \text{if } t-1 < T_x \leq t \\ 0 & \text{otherwise} \end{cases}$, ${}_{\xi}J_x = \begin{cases} 0 & \text{if } 0 < T_x \leq \xi \\ e^{-\Delta(t)} & T_x \geq \xi \end{cases}$.
In the previous expression, T_x is a random variable which represents the remaining lifetime of an insured aged x, $\Delta(t) = \int_0^t r_u du$ is the accumulation function of the spot rate.

3.1 Financial and mortality scenario

The valuation of the financial instruments involving the policy will be made assuming a two factor diffusion process obtained by joining the Cox-Ingersoll-Ross (CIR) model for the interest rate risk and a Black-Scholes (BS) model for the stock market risk; the two sources of uncertainty are correlated. In general, interest rate dynamics $r_t; t = 1, 2, \ldots$ and described by means of the diffusion process

$$dS_t = f^r(r_t, t)dt + l^r(r_t, t)dZ_t^r\,, \tag{3}$$

where $f^r(r_t, t)$ is the drift of the process, $l^r(r_t, t)$ is the diffusion coefficient and Z_t^r is a Standard Brownian Motion.

Clearly, on the fair pricing of our policy, it is very important to specify the reference portfolio dynamics. The diffusion process for these dynamics is given by the stochastic differential equation

$$dS_t = f^S(S_t, t)dt + g^S(S_t, t)dZ_t^S\,, \tag{4}$$

where S_t denotes the price at time t of the reference portfolio and Z_t^S is a Standard Brownian Motion with the property $Cov(dZ_t^r, dZ_t^S) = \varphi dt \quad \varphi \in R$.

For the dynamics of the process $\{\mu_{x+t:t}; t = 1, 2, \ldots\}$, we propose a model based on the Lee-Carter methodology. A widely used actuarial model for projecting mortality rates is the reduction factor model for which

$$\mu_{y:t} = \mu_{y:0} RF(y, t), \tag{5}$$

subject to $RF(y, 0) = 1 \; \forall y$, where $\mu_{y:0}$ is the mortality intensity of a person aged y in the base year 0, $\mu_{y:t}$ is the mortality intensity for a person attaining age y in the future year t, and the reduction factor is the ratio of the mortality intensity. It is possible to target RF, in a Lee-Carter approach, $\mu_{y:0}$ being completely specified. Thus, $\mu_{y:0}$ is estimated as $\widehat{\mu}_{y:0} = \sum_t d_{y:t} / \sum_t e_{y:t}$ where $d_{y:t}$ denotes the number of deaths at age y and time t and indicates the matching person years of exposure to the risk of death. Taking the logarithm of equation (3) we have $\log \mu_{y:t} = \log \mu_{y:0} + \log RF(y, t)$ s.c. $\log RF(y, 0) = 0$. Defining $\alpha_y = \log(\mu_{y:0})$ and $\log\{RF(y, t)\} = \beta_y k_t$ the Lee-Carter structure is reproduced.

In fact the Lee-Carter model for death rates is given by $\ln(m_{yt}) = \alpha_y + \beta_y k_t + \epsilon_{yt}$ where m_{yt} denotes the central mortality rates for age y at time t, α_y describes the shape of the age profile averaged over time, k_t is an index of the general level of mortality while β_y describes the tendency of mortality at age y to change when the general level of mortality k_t changes. ϵ_{yt} denotes the error. In this framework, for our purposes, with $y = x + t$, one can use the following model for the time evolution of the hazard rate $\mu_{x+t:t} = \mu_{x+t:0} e^{\beta_{x+t} k_t}$.

4 Numerical Proxies for the Quantile Reserve via Simulation Procedures

4.1 The problem background

In this section we present a simulation procedure to calculate the quantile reserve, providing a practical application of the mathematical and accounting tools presented previously. In particular, we will quantify the two critical values of the quantile reserve $W^*_{0.95}(t)$ $W^*_{0.99}(t)$. The calculation of the quantile reserve values requires the knowledge of the distribution of $W(t)$. To this aim, we use a Monte Carlo simulation procedure which, as well known, is typically employed to model random processes that are too complex to be solved by analytical methods.

As a first step, we develop the statement of the problem giving the mathematical relationship between the input and output variables. On the basis of the model presented in Section 3, the output is given by the financial position of the insurer at time t, $W(t)$, and the input variables are given by the time of valuation t, the survival probabilities and the term structure of interest rates. In this order of ideas, we assume that the best prediction for the time evolution of the surviving phenomenon is represented by a varying fixed set of survival probabilities, opportunely estimated taking into account the improving trend of mortality rates. As a consequence, in our application, the first two inputs are deterministic while the random input is represented by the model describing interest rate distribution which is essential in a fair valuation context.

We refer to the following hypothesis: for the interest rates scenario we use a CIR model specifying the drift function and the diffusion coefficient in formula (3) as follows: $f^r(r_t, t) = k(\theta - r_t)$, $l^r(r_t, t) = \sigma_r\sqrt{r_t}$ where k is the mean reverting coefficient, θ the long term period "normal" rate and σ_r the spot rate volatility. We assume for the CIR process $k = 0.0452, \theta = 0.026$ and $\sigma = 0.00053$ and the initial value $r_0 = 0.0172$. For the reference portfolio dynamics, we assume a BS type model and referring to formula (4): $f^S(S_t, t) = \mu_S S_t$ and $g^S(S_t, t) = \sigma_S S_t$ where μ_S is the continuously compounded market rate, assumed to be deterministic and constant, and σ_S is the constant volatility parameter. In our application $\mu_S = 0.03$ and $\sigma_S = 0.20$. For the mortality, we refer to three different scenarios:

a) The survival probabilities deduced by Italian data for the period 1947-1999 using the Lee-Carter methodology.
b) The survival probabilities deduced by Italian Male Mortality called RG48.
c) The survival probabilities deduced by Italian Male Mortality called SIM92.

The example of application we propose is referred to a life insurance participating contract. In particular, we quantify, at the beginning of the contract, the two critical values $R^*_{0.95}(t)$ and $R^*_{0.99}(t)$ of the reserve distribution. The output of the simulation procedure is a sample which gives N values for $W(t)$, N being the number of simulations. In order to perform the simulation procedure, it is necessary to obtain the discrete time equation for the chosen SDE describing the evolution in time of the interest rates. We choose the first order Euler's approximation scheme, obtaining the following sample path simulation equation:

$$r_{k\Delta t} = r_{(k-1)\Delta t} + \alpha(\mu - r_{(k-1)\Delta t}) + \sigma\sqrt{r_{(k-1)\Delta t}\,\Delta t} \cdot \epsilon_k \qquad k = 1, 2, \ldots, T\,, \quad (6)$$

where $\{\epsilon_k \approx N(0, 1)\}$.

This approximation scheme is characterized by an easy implementation and a simple interpretation of the results. The discretized process we consider can be represented by the sequence $\{r_{\Delta t}, r_{2\Delta t}, \ldots, r_{k\Delta t}\}$, where k is the number of time steps, Δt is a constant and T is the time horizon.

The following simulation procedure is carried out in order to gain a sample of N values of $W(t)$:

a) Generation of T pseudo-random values $\{\epsilon_k \approx N(0, 1)\}$.
b) Computation of one simulated path for the stochastic interest rate $\{r_{k\Delta t}\}$ using the T values obtained in step (a).
c) Computation of one value of the reserve on the basis of the previous results.

The simulation procedure will be repeated N times to gain N values for $R(t)$. At this point our purpose is to quantify the two critical values of the reserve distribution $W^*_{0.95}(t)$ and $W^*_{0.99}(t)$. As reported in the previous sections, since the reserve is a liability, we are interested in the right hand tail of the distribution.

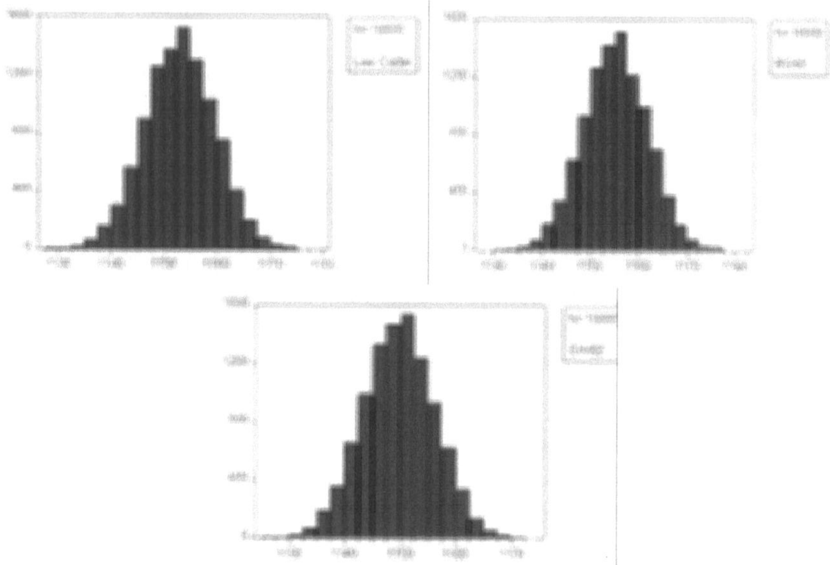

Fig. 1. Empirical distribution of W(t).

4.2 Numerical results

The numerical example we propose refers to a portfolio life participating contract issued to an individual aged 40 with time to maturity 20 years.

Table 1 confirms the asymptotic behavior of the empirical distribution of the random variable $R(t)$. Now, as we already recalled, the Glivenko-Cantelli theorem is verified, in the sense that, as we can easily observe, $R(t)$ approximates a normal distribution. In particular, kurtosis has a value of about 3 and skewness has a value of about 0. It is well known that a normal variable has a kurtosis of 3 and a skewness equal to zero, therefore the obtained values can be considered acceptable. Moreover, we have conducted the Jarque Bera test. As well known, the J-B is a statistic for

Table 1. W empirical distribution. N = 10000.

	LC	RG48	SIM92
M[W]	1153.047	1155.166	1149.773
Max	1178.297	1180.542	1174.329
Min	1128.005	1129.996	1125.409
Skewness	−0.012903	−0.012904	−0.012915
Kurtosis	3.009453	3.009717	3.008749
JB	0.314699	0.316859	0.309875
Probability	0.854406	0.853483	0.856469

Table 2. Initial quantile reserve values.

	LC	RG48	SIM92
W*(0.05)	1163.829	1166.009	1160.239
W*(0.01)	1168.116	1170.307	1164.439

testing whether the series is normally distributed. The J-B test is known to have very good properties in testing for normality; it is easy to calculate and it is commonly used in econometrics. In our case, as the probability is equal to about 0.85, we can accept the null hypothesis of the normal distribution of $R(t)$.

In Table 2, we have calculated the initial values of the quantile reserve at a confidence level of 95% and 99%. As one can see, we obtain, in each mortality scenario, values higher than expected. The difference between W* values and M[W] values can be interpreted as an absolute index of the riskiness borne out by the insurer due to the uncertainty regarding interest and mortality rates. Finally, concerning the influence of the mortality factor, in our application, if we perform a comparison between the quantile reserve in each mortality scenario, one can see that the difference between W^* and M[W] is quite stable, and the quantile values are quite similar.

References

[Bac01] Bacinello, A. R.: Fair Pricing of Life Insurance Participating policies with a minimum interest rare guaranteed. Astin Bulletin, 275–297 (2001)

[BH06] Ballotta, L., Haberman S.: The Fair Valuation problem of guaranteed annuity option: the stochastic mortality case. Insurance:Mathematics and Economics, Vol. **38**, N. 1, 195–214 (2006)

[BJ1980] Bera, A., Jarque C.: Efficient Test for normality, heteroskedasticity and serial independence of regression residuals: Monte Carlo evidence. Economic Letter, Vol. **6**, 255–259 (1980)

[BS1973] Black F., Scholes M.: The pricing of option and corporate liabilities. Journal of Political Economy, 637–654 (1973)

[BDG04] Bin Dog L., Gils D. E. A.: An Empirical likelihood ratio test for normality. Econometrics Working Papers EWP0401 University of Victoria (2004)

[Bu04] Buhlmann H.: Multidimensional Valuation. *Working Paper* (2004)

[CIR1985] Cox J., Ingersoll J, Ross S.: A theory of the term structure of interest rates. Econometrica, 385–408 (1985)

[IAIS] IAIS, January. International Accounting Standard Board: International Financial Reporting Standard 4 Insurance Contracts (2004)

[L00] Lee R.: The Lee Carter method of forecasting mortality with various extension and applications. North American Actuarial Journal, 80–93 (2004)

[MP01] Milevsky M.A., Promislow S.D.: Mortality Derivatives and the option to annuities. Insurance: Mathematics and Economics, 299–318 (2001)

[RH03] Renshaw A.E., Haberman S.: On the forecasting of mortality reduction factor. Insurance: Mathematics and Economics, 379–401. (2003)

Classifying Italian Pension Funds via GARCH Distance

Edoardo Otranto and Alessandro Trudda

Summary. The adoption of pension funds in the Italian social security policy has increased the offer of several investment funds. Workers have to decide what kind of investment to perform, the funds having a different composition and a subsequently different degree of risk. In this paper we propose the use of a distance between GARCH models as a measure of different structure of volatility of some funds, with the purpose of classifying a set of funds. Furthermore we extend the idea of equivalence between ARMA models to the GARCH case to verify the equality of the risk of each couple of funds. An application on thirteen Italian funds and fund indices is performed.

Key words: Agglomerative algorithm; Cluster analysis; GARCH models; Pension funds; Risk profile.

1 Introduction

The main aim of a pension fund is to raise workers' savings and to invest them according to an accurate policy of asset allocation, in order to give back the hoarded capital as a life annuity. Therefore, the most important index pension fund refers to the global asset return since it influences both periodic contributions and future benefits. For this reason, the valuation of a pension fund is often related to the performance rather than to the risk level of asset portfolio. According to [RF03], the shortfalls of the USA pension after 2001 are not the consequence of a poor market performance, but the inevitable result of actuarial and fiscal accounting practices. In fact, in the long term, the expected asset growth must be higher for equities than for bonds. [Bad03] takes an opposite position and argues that risky investment like equity has no place in pension fund portfolios. Such a rule should be applied particularly for pension funds when they concern public and not complementary pension. [McC06] discusses a "moral hazard hypothesis"; in particular underlining how pension funds assume "superfluous risk" looking for higher returns without an adequate adjust for risk.

In Italy, the pension plans of private sector workers are run by private funds controlled by the government. [Tru05] studies actuarial balances and dynamics for

Cira Perna, Marilena Sibillo (Eds.): Mathematical and Statistical Methods in Insurance and Finance

Italian independent consultant pension (ICP) plans. The analysis of each random variable shows that a marginal increase of global asset return gives an important reduction of default probability. In his analysis, [Tru05] presents evidence that this kind of fund belongs to the first pillar pensions group and they cannot be assimilated as complementary pension fund random variables due to technical and social reasons. Therefore, the global asset return analysis should refer to a risk benchmark rather than a return benchmark, recognizing the maximum risk profile. The performance of these ICP funds is variegated depending on their asset composition. Some of them have only real or bond investment while others have a high stock market component. The 2006 annual report on the condition of public and private Italian welfare Agency, affirms that most of these funds are now shifting to higher risk level portfolios. The Agency stresses the social rather than speculative function of such funds which valuate the maximum risk profile in function of the liability for pension payments. On the other hand, investors would like to choose from funds with similar behavior but different degrees of risk. For these reasons, an important task is to classify the different funds with different degrees of risk with respect to the dynamics of their movements. In statistical terms, we have to follow the evolution of the time series referring to the single fund and to perform an appropriate cluster analysis. In this paper we use an agglomerative algorithm proposed by [Otr04] based on the distance between two time series following GARCH models ([Bol86]). This distance is an extension of the well known distance between invertible ARMA models proposed by [Pic90]. The logic of this approach in this framework is explained by the link existing between volatility and investment risk. High volatility periods correspond to turmoil phases in the dynamics of the time series studied; if we represent the conditional variance of the series with GARCH models, the GARCH structure will reflect the behavior of the volatility. In other words, similar GARCH models represent similar volatility behavior and similar investment risks. If we base the analysis only on the dynamic parts of the GARCH models (excluding the constant term), we can compare the series with respect to their evolution; in other words, series with different risk (different degree of volatility) can have similar dynamics. For this purpose we need a benchmark time series; a logical choice could be a fund which potentially presents the most or the least hazardous global asset risk profile.

The procedure we propose provides three products: 1) a classification of the funds in clusters having similar dynamics; 2) the evaluation of the series having equal risk in the same cluster; 3) the comparison of all the series, independently of the corresponding cluster, to detect those having similar risk.

All the results are derived from a Wald type test.

In the next section we briefly recall the GARCH distance proposed by [Otr04] and the agglomerative algorithm used; in Section 3 we show an example of classification based on this methodology using a risk benchmark to mark a maximum risk profile as suggested by [Tru05]. Some final remarks follow.

2 Distance Between GARCH Models

Let us consider two time series following the models ($t = 1, ..., T$):

$$\begin{aligned} y_{1,t} &= \mu_1 + \varepsilon_{1,t} \\ y_{2,t} &= \mu_2 + \varepsilon_{2,t}, \end{aligned} \tag{1}$$

where $\varepsilon_{1,t}$ and $\varepsilon_{2,t}$ are disturbances with mean zero and time-varying variances. We suppose that the conditional variances $h_{1,t}$ and $h_{2,t}$ follow two different and independent GARCH(1,1) structures:[1]

$$\begin{array}{l} Var(y_{1,t}|I_{1,t-1}) = h_{1,t} = \gamma_1 + \alpha_1\varepsilon^2_{1,t-1} + \beta_1 h_{1,t-1} \\ Var(y_{2,t}|I_{2,t-1}) = h_{2,t} = \gamma_2 + \alpha_2\varepsilon^2_{2,t-1} + \beta_2 h_{2,t-1} \end{array}, \tag{2}$$

where $I_{1,t}$ and $I_{2,t}$ represent the information available at time t and $\gamma_i > 0$, $0 < \alpha_i < 1$, $0 < \beta_i < 1$, $(\alpha_i + \beta_i) < 1$ $(i = 1, 2)$.

As well known, the squared disturbances in (2) follow ARMA(1,1) processes:

$$\varepsilon^2_{i,t} = \gamma_i + (\alpha_i + \beta_i)\,\varepsilon^2_{i,t-1} - \beta_i\left(\varepsilon^2_{i,t-1} - h_{i,t-1}\right) + \left(\varepsilon^2_{i,t} - h_{i,t}\right), \quad i = 1, 2, \tag{3}$$

where $\varepsilon^2_{i,t} - h_{i,t}$ are mean zero errors, uncorrelated with past information. The two GARCH(1,1) models can be compared in terms of the distance measure proposed by [Pic90]. In fact, substituting into (3) the errors with their ARMA(1,1) expression, we obtain the AR(∞)representation:

$$\varepsilon^2_{i,t} = \frac{\gamma_i}{1-\beta_i} + \alpha_i \sum_{j=1}^{\infty} \beta_i^{j-1} \varepsilon^2_{i,t-j} + \left(\varepsilon^2_{i,t} - h_{i,t}\right). \tag{4}$$

The general form of the [Pic90] distance is:

$$\left[\sum_{j=1}^{\infty} (\pi_{1j} - \pi_{2j})^2\right]^{1/2}, \tag{5}$$

where π_{1j} and π_{2j} are the autoregressive coefficients of two AR processes. Using (4), we can express the distance between two GARCH(1,1) models as ([Otr04]):

$$d = \left[\sum_{j=0}^{\infty} \left(\alpha_1\beta_1^j - \alpha_2\beta_2^j\right)^2\right]^{1/2}.$$

---

[1] In our applications, we have noted that the $GARCH(1,1)$ models fit the series analyzed, so we explain the idea of distance only for this case. Note that [BCK92] underlines that the GARCH(1,1) model fits excellently a wide range of financial data. For an extension to a generical $GARCH(p,q)$ model, see [Otr04].

Developing the expression in square brackets:

$$d = \left[\alpha_1^2 \sum_{j=0}^{\infty} \beta_1^{2j} + \alpha_2^2 \sum_{j=0}^{\infty} \beta_2^{2j} - 2\alpha_1\alpha_2 \sum_{j=0}^{\infty} (\beta_1\beta_2)^j \right]^{1/2} = \tag{6}$$

$$= \left[\frac{\alpha_1^2}{1-\beta_1^2} + \frac{\alpha_2^2}{1-\beta_2^2} - \frac{2\alpha_1\alpha_2}{1-\beta_1\beta_2} \right]^{1/2}.$$

Note that in the previous developments, the constant $\gamma_i/(1-\beta_i)$ was not considered; in effect, it does not affect the dynamics of the volatility of the two series, expressed by the autoregressive terms. This will be an important task when we use this distance measure to classify the time series; in fact, we are measuring the similarity of the dynamics of the volatility of two time series and not the similarity of their volatility.

The hypothesis of null distance can easily be tested noting that (6) is equal to zero if and only if the following hypotheses are verified:

$$\begin{aligned} \alpha_1 &= \alpha_2 \\ \beta_1 &= \beta_2. \end{aligned} \tag{7}$$

These hypotheses are contemporaneously verified using the Wald statistic:

$$W = (\mathbf{A}\widehat{\boldsymbol{\theta}})'(\mathbf{A}\widehat{\boldsymbol{\Lambda}}\mathbf{A}')^{-1}(\mathbf{A}\widehat{\boldsymbol{\theta}}), \tag{8}$$

where $\widehat{\boldsymbol{\theta}}$ represents the maximum likelihood estimator of $\boldsymbol{\theta} = (\alpha_1, \beta_1, \alpha_2, \beta_2)'$, whereas $\mathbf{A} = [\mathbf{I}_2, -\mathbf{I}_2]$, where $\mathbf{I}_2$ is the 2×2 identity matrix. $\widehat{\boldsymbol{\Lambda}}$ is the estimated covariance matrix of $\widehat{\boldsymbol{\theta}}$; estimating the GARCH models independently, $\widehat{\boldsymbol{\Lambda}}$ is a block diagonal matrix, with the [BW92] robust covariance matrix of the parameters of each model in each block.

Statistic (8) follows the chi-square distribution with 2 degrees of freedom.

This distance can be used to cluster n time series in homogenous groups, having similar GARCH structure. In other words, we insert the series with no significant distance in the same cluster. For this purpose, an agglomerative algorithm could be used; following [Otr04] it can be developed in the following steps:

1. Choose an initial benchmark series.
2. Insert into the group of the benchmark series, all the series with a distance not significantly different from zero (using statistic (8)).
3. Select the series with the minimum distance from the benchmark significantly different from zero; this series will be the new benchmark.
4. Insert in the second group, all the remaining series with a distance from the new benchmark that is not significantly different from zero.
5. Repeat steps 3 and 4 until no series remain.

Note that the number of groups is not fixed a priori or chosen after the clustering, but derives automatically from the algorithm.

Clearly, to classify the series we need a starting point, in the sense that the result will be different, changing the series adopted as the initial benchmark. For our purposes, we choose the stock index fund having the highest potential risk or that with minimum risk, formed by the Italian government Z.C. bond index.

As stated in the introduction, this classification does not refer to clusters containing series with similar risk, but series with similar dynamics. This idea is similar to the concept of parallelism between ARMA models, introduced by [SW85]. In other words, we are grouping series with a similar dynamic part and, as a consequence, a similar predictability ([OT07]). To verify if the series have the same volatility (which is equivalent to assuming that they have the same risk), we have to test the hypothesis that the two GARCH models are the same. This idea corresponds to another concept introduced by [SW85] for ARMA models, which is equivalence. We affirm that two GARCH processes are equivalent if the following hypotheses are verified:

$$\begin{aligned} \gamma_1 &= \gamma_2 \\ \alpha_1 &= \alpha_2 \\ \beta_1 &= \beta_2. \end{aligned} \tag{9}$$

The test used to verify this hypothesis is again (8), in which we have only to modify in an appropriate way the dimension of the matrices **A** and Λ.

3 A Classification of Funds

We consider thirteen time series from November 1995 to December 2000 (daily data; 1290 observations; sources: Independent Consultants Pension Funds (ICPF) balance sheet, Complementary Pension Funds (CPF) balance sheet, and Supplements from the Bank of Italy Statistical Bulletin) relative to some basis and complementary Italian pension funds and fund indices. The asset portfolio composition of these funds is the following:

A = CPF with mixed (bond and stock funds) financial products.
B = Z.C. Italian government bonds index.
C = CPF with real estate investments, 10 year duration benchmark bonds and small stock component.
D = ICPF with real estate and low year duration benchmark bonds.
E = ICPF with liquidity, properties and 30 year duration benchmark bonds.
F = CPF with 3 year duration benchmark bonds and small stock component.
G = ICPF with prevalence of investments in government bond funds.
H = Investment fund with interest rate indexation benchmark.
I = ICPF with prevalence of bond investments.
J = Italian stock funds index.
K = CPF with prevalence of bond fund investments.
L = Italian total funds index.

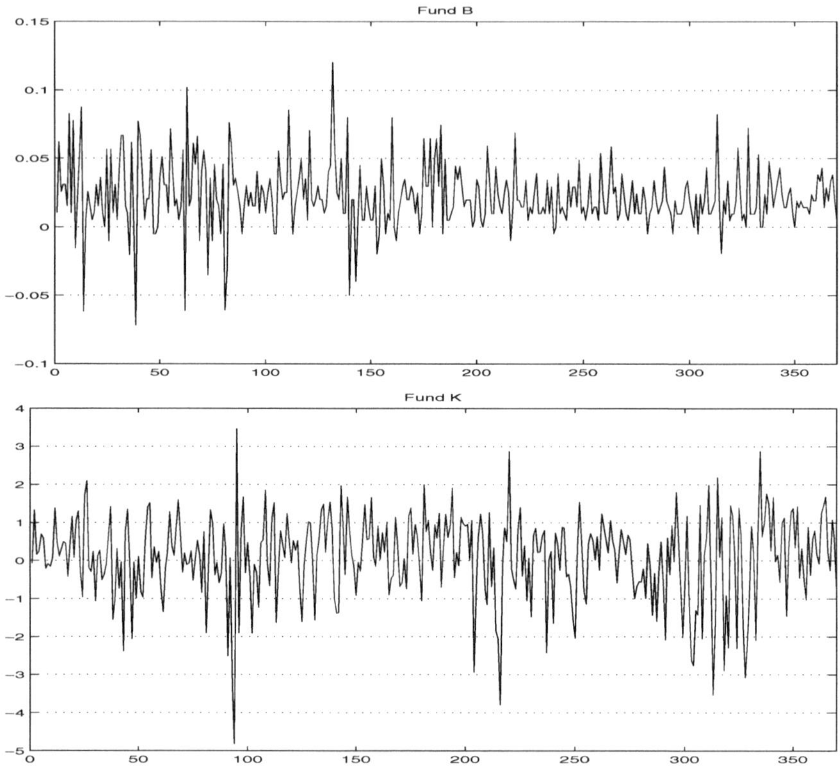

Fig. 1. Returns of fund indices B and K.

We can note that the possibility of choice for investors is large, providing funds with different degrees of risk. In practice, the choice varies from funds index B, which shows a poor degree of volatility, because it invests mostly in properties, real estate and bond funds, maintaining a minimum return with high tendency on stock market investments, to fund index K, which has a total component on stock market investments. In Fig. 1, the returns (the first differences of the logs of the series) of the last 370 observations are compared, showing these characteristics. Note as fund index B shows small returns prevalently positive, assuring a minimum gain to the investor; fund index K has large positive and negative oscillations, denoting the possibility of large gains but also large risks.

The investor's choice will depend essentially on the inclination to the risk and the historical behavior of the series. In other words, a good criterion of choice could be to select the series which follow similar patterns and then to evaluate the level of risk of the group selected.

The procedure we propose provides a classification of the funds based on their dynamics; this is obtained by applying the algorithm proposed in the previous section. Furthermore, we can detect, in the same group, if the series have a similar degree of risk; this is obtained using the hypothesis (9). For each time series we have estimated a GARCH(1,1) model as (2).[2] Assuming K, which is theoretically the most risky index, as initial benchmark, we obtain the distance (6) from K shown in Table 1.

Table 1. GARCH(1,1) distance of twelve funds from fund index K.

A	B	C	D	E	F	G	H	I	J	L	M
0.013	0.167	0.093	0.131	0.217	0.064	0.122	0.156	0.111	0.114	0.091	0.016

Applying the agglomerative algorithm,[3] we obtain the clusters shown in Table 2:

Table 2. Clusters obtained using fund index K as a benchmark.

CLUSTER 1	CLUSTER 2	CLUSTER 3	CLUSTER 4
K A M	F C J G D L	H	I B E

It is interesting to note that the same classification is obtained using the fund with the lower risk profile as a benchmark, (fund index B). Shown in Table 3 are the distances from it. Note that the distances from B are in general larger than the distances from K (apart from E and I), but this indicates only that these three series have a peculiar behavior, different from all the others. In other words, this classification seems robust enough and the series belonging to the same cluster would have similar dynamics.

Table 3. GARCH(1,1) distance of twelve funds from fund index B.

A	C	D	E	F	G	H	I	J	K	L	M
0.176	0.257	0.293	0.096	0.226	0.283	0.318	0.083	0.276	0.167	0.235	0.156

Given this classification, we can verify the three hypotheses in (9). In practice, we verify these hypotheses for each couple of series belonging to the same cluster (apart from cluster 3 which contains only one series). The result is that only the couple F and C, and the couple J and D can be considered to have the same volatility at a nominal size of the test equal to 5%.

If the investors are interested only in detecting the funds with a similar risk, we can verify the hypotheses (9) for each couple of series. Besides the two cases belonging to the same cluster, we obtain that we do not reject the null hypothesis of

[2] The estimation results are not shown to save space. They are available on request.

[3] Test (8) is performed with a nominal size of 5%.

equal risk also for the couple D and H, and the couple E and K. In particular, this last result is interesting; in practice we obtain that fund E, constituted by liquidity, properties and bonds, has the same risk as the most risky fund; at the same time, the two funds follow different dynamics, with E which has an evolution similar to that of the less risky fund index B. Cluster 4 is a clear demonstration that series with similar dynamics can show a very different risk.

4 Concluding Remarks

The stock market deadline, after 2001, opened a discussion about the risk of structure pension fund investment. Pension corporations often increase asset allocation risk to obtain higher values of the expected global asset return. Many authors show evidence that this is not a correct strategy, in particular for the first pillar case, because social function prevails on the speculative function. Some tests on USA pension investment provide evidence of a "moral hazard" problem because accounting rules incentives pension corporations to assume "superfluous risk" without adequately adjusting for risk. In Italy there are private institutions, controlled by government, that manage the bases of pension plans independently of consultants. The 2006 annual report of the welfare Government Agency disapproves of the tendency of these institutions to shift to higher risk level portfolios and in some cases suggest a return to maintaining a prudence profile. A large part of the analysis values the asset allocation strategies in qualitative terms without an accurate analysis of global asset return dynamics.

In this paper we have introduced a statistical procedure to classify pension funds in function of two different components of financial risk: the dynamics and the volatility of global asset returns. Equal dynamics can be interpreted as equal predictability of the volatility, whereas equal volatility can be seen as similar risk level associated with the returns on investment. The idea is that the GARCH structure of the global asset return series incorporates both the components, so that we can use the GARCH distance of [Otr04] to classify the funds and a set of Wald type tests to verify the equal volatility of each couple of series.

The use of the distance implies the adoption of a benchmark series to start the classification; in this paper we have used the fund index considered as the most risky among the series available and one considered of minimum risk. Of course the choice of the benchmark depends on the purpose of the analysis and could refer to a hypothetical GARCH model. For example, we could choose to compare the series with a hypothetical series with constant variance ($\alpha_1 = 0$ and $\beta_1 = 0$ in (2)) or with a hypothetical series without risk (in this case also $\gamma_1 = 0$). In this case, the distance measures how the series analyzed is different from a series without dynamics.

References

[Bad03] Bader, L. N.: The case against stock in corporate pension funds. Society of Actuaries Newsletter of the Pension Section, **51**, 17–19 (2003)

[Bol86] Bollerslev, T.: Generalized autoregressive conditional heteroskedasticity. Journal of Econometrics, **31**, 307–327 (1986)

[BCK92] Bollerslev, T., Chou, R. Y., Kroner K. F.: ARCH modeling in finance: A review of the theory and empirical evidence. Journal of Econometrics, **52**, 5–59 (1992)

[BW92] Bollerslev, T., Woolridge, J. M.: Quasi-maximum Likelihood Estimation and Inference in Dynamic Models with Time-varying Covariances. Econometric Reviews, **11**, 143–172 (1992)

[McC06] McClurken: Testing a Moral Hazard Hypothesis in Pension Investment. Washington University, Washington (2006)

[Otr04] Otranto, E.: Classifying the markets volatility with ARMA distance measures. Quaderni di Statistica, **6**, 1–19 (2004)

[OT07] Otranto, E., Triacca, U.: Testing for Equal Predictability of Stationary ARMA Processes. Journal of Applied Statistics, forthcoming (2007)

[Pic90] Piccolo, D.: A distance measure for classifying ARIMA models. Journal of Time Series Analysis, **11**, 153–164 (1990)

[RF03] Ryan, R. J., Fabozzi, F. J.: The pension crisis revealed. The Journal of Investing, **12**, 43–48 (2003)

[SW85] Steece, B., Wood, S.: A test for the equivalence of *k* ARMA models. Empirical Economics, **10**, 1–11 (1985)

[Tru05] Trudda, A.: Casse di previdenza: Analisi delle dinamiche attuariali. Giappichelli, Torino (2005)

The Analysis of Extreme Events – Some Forecasting Approaches*

Massimo Salzano

Summary. Policy makers often have to deal with extreme events. We consider the two main approaches for the analysis of these events, statistical and physical statistics, which are usually considered dichotomically. We quickly delineate both limits and advantages of their tools and some new tentatives for overcoming the historical limits of the latter kind of approach. We highlight that for dealing adequately with extreme events it seems necessary to make use of a new kind of Decision Theory and new tools.

Key words: Extreme events; Extreme Value Theory (EVT); Self-organized Criticality (SOC).

1 Introduction

In many diverse areas of application, the extreme values of a process are of primary importance. Large natural catastrophes, the failure of engineering structures, regional power blackouts, traffic gridlock, disease and epidemics, etc. exhibit this type of characteristic. Policy makers often have to deal with extreme events (XE). Therefore, the tools that must be used for their forecasting are of interest both for researchers and for practitioners. Intense attention and efforts are devoted in the academic community, in government agencies, and in the industries that are sensitive to or directly interested in these risks, to the understanding, assessment, mitigation, and if possible, prediction of these events. Different types of XE pose different challenges to decision making. Therefore, we have to differentiate between two kind of events: Rapid (like earthquakes) or Creeping (like global warming)[2]. According to the literature, these sort of events could be both exogenously (closed systems[3]) or endogenously originated (open systems)[4]. A number of different techniques can be justifiably employed to analyze the extremes of one or more processes. We could make a dis-

* This research is supported by a grant from Miur-Firb RBAU01B49F/001

[2] Stewart, Bostrom (2002).

[3] Closed Systems ("in which all the components of the system are established independently and are known to be correct".); Open Systems ("in which all the components of the system are not established independently or are not known to be correct").

[4] Pielke (2001).

Cira Perna, Marilena Sibillo (Eds.): Mathematical and Statistical Methods in Insurance and Finance

tinction between methods of statistical description vs. methods highlighting, their causes and predictability. Whilst for the practitioner it suffices to apply some simple statistical description, or forecasting[5], for the researcher and for the "policy maker" the emphasis must be on the causes and predictability of patterns within crashes. Even if it is only the second kind of systems which present "deep uncertainty"[6], important decisions based on methods adequate for the first kind are often clouded by inherent uncertainty, and in many instances, efforts to reduce uncertainty have the opposite effect (Pielke 2001). Efforts to reduce uncertainty can lead to discovery of the vast complexities associated with phenomena that evolve slowly over long periods[7]. The different approaches are usually considered in a dichotomic way. We will try to consider here the different hints they could offer to researchers. We begin with a brief review of the classic statistical approach – EVT – and its limits, and then we consider a different approach proposed on the basis of self-organized criticality (SOC). We conclude by considering some implications of SOC for dealing with risk. Bankes (1993) highlighted two kinds of quantitative model: a) A "consolidative" model, (i.e. a controlled laboratory experiment; the weather forecast and engineering design models) that considers the system as being closed. It aims to include all relevant facts in a single package and use the resulting system as a proxy for the actual system. These kinds of models could be particularly relevant to engineering decision making. b) An exploratory model that is one in which all components of the system being modeled are not established independently (interdependence) or are not known to be correct. In such a case, there is a "deep uncertainty" and the model allows for experiments to investigate the consequences of various assumptions, hypotheses, and uncertainties.... (Bankes 1993) They can: shed light on the existence of unexpected properties, facilitate hypothesis generation and identify limiting, worst-case, or special scenarios. The two kinds of quantitative model developed different kinds of approaches. Mainly: a) Statistical Extreme Value Theory[8]and b) Extremes and Statistical Mechanics. For the former, based on Gaussian or its derived distributions, XE are very improbable events reacting to exogenous causes, while for the latter, they depend on the internal structure of the system. The self-organizing capacity of the system gives rise to a power law scaling called Self-organized Criticality (SOC). A strong difference between the two approaches regards the fact that the latter tries to explain more the mechanism at the base of their insurgence while the former tends to forecast more[9]. The statistics of SOC being non-Gaussian. This is very much work in progress, but seems to be a potentially fruitful line of research. The approach is to build from the bottom up. These models tend to be viewed as "exotic" by statistic literature because of their unusual properties (e.g., infinite moments). An alternate view is based on mathematical, statistical, and data-analytic arguments and suggests that scaling distributions

[5] "Whilst for the practitioner it suffices to apply extreme value theory, a simple statistical description, here the emphasis is on the causes and ..." Adventures on the edge of chaos – www.iop.org/EJ/abstract/1469-7688/4/2/R01.

[6] Bankes (1993).

[7] Pielke (2001).

[8] For a good survey of approaches regarding the risk of extremes see: Bier and al. (1999).

[9] An example of the limits of the classical model can be found in Feigenbaum, J. (2003) that says: "While the RBC model is useful for explaining the correlations between macroeconomic variables over the course of the business cycle, it does not address why these variables fluctuate in the first place. The reason why these fluctuations are so mysterious can be explained using statistical mechanics".

should be viewed as "more normal than Normal" as has been advocated by Mandelbrot and others[10].

2 Kinds of Approaches for the Analysis of Extreme Events

Many kinds of possible approaches are summarized in Bier and al. (1999). The most used approach seems to be the Extreme Value Theory.

2.1 Statistical extreme value theory

Extreme Value Theory (EVT) is an established area in probability theory and mathematical statistics. It deals with "extreme deviations from the median of probability distributions"[11], of finite time series provided by a very large collection of random observations, assumed to be independent and identically distributed (i.i.d.), and tries to assess the type of probability distributions generated by processes. Gumbel (1958) showed that for any well-behaved initial distribution (i.e., F(x) is continuous and has an inverse), only a few models are needed, depending on whether you are interested in the maximum or the minimum, and also if the observations have upper or lower boundaries. It is used for assessing risk for highly unusual events. Two main variants exist according to which data is considered:

The basic theory approach

This approach[12] considers all the data[13]. Its main tenet[14] is that if $(Xn)n \in N$ is a sequence of i.i.d. real-valued random variables, and if the *maximum* $Yn \equiv max(X1, \ldots, Xn)$ suitably "standardized", has a limit distribution of n, then this distribution belongs to a generalized extreme value distribution [15](Von Mises): $P\gamma(x) = exp\left(-\left(1+\gamma\frac{x-m}{\sigma}x\right)_+^{-1/\gamma}\right), \quad x \in R.$ From it we can obtain the standard three extreme value distributions: a) if $a+ = max(0, a)$, so that $\gamma > 0$ we obtain the Freéchet distribution; b) if $\gamma < 0$ we obtain the Weibull distribution, and c) since $(1 + 0 \cdot x)^{-1/0} = e^{-x}$ the value $\gamma = 0$ corresponds to the Gumbel distribution.

a) (Fréchet) $P_{Fr}(x) = \begin{cases} 0, & x \le 0, \\ e^{-x^{-\alpha}}, & \alpha > 0, \ x > 0; \end{cases}$

[10] For some relevant results from probability theory and illustration of a powerful statistical approach for deciding whether the variability associated with observed event sizes is consistent with an underlying Gaussian-type (finite variance) or scaling-type (infinite variance) distribution, see: Willinger and al.(2004).

[11] For a simple survey, see: http://en.wikipedia.org/wiki/Extreme value theory

[12] Burry K.V. (1975). For a review, see: Kotz; Nadarajah (1999) – I Chap.

[13] This approach conforms to the first theorem in EVT; see: Fisher and Tippett (1928), and Gnedenko (1943).

[14] Albeverio and al.

[15] Einmahl H.J., de Haan L., There are many distributions which do not belong to the three mentioned domains of attraction. Nevertheless, most applied distributions (Pareto-like distributions (Cauchy), normal beta, arcsin, ...) do.

b) (Gumbel) $P_{Gu}(x) = e^{-e^{-x}}, \quad x \in R;$

c) (Weibull) $P_{Wei}(x) = \begin{cases} e^{-(-x)^{-\alpha}}, & \alpha > 0,\ x \leq 0, \\ 1, & x > 0 \end{cases};$

where $a+ = max(0, a)$, so that $\gamma > 0$ corresponds to the Fréchet distribution, $\gamma < 0$ corresponds to the Weibull distribution, and since $(1 + 0 \cdot x)^{-1/0} = e^{-x}$ the value $\gamma = 0$ corresponds to the Gumbel distribution.

The parameter γ is called the extremum value index, and it provides important information about the tail of the underlying distribution $P_\gamma.t$. Often, when there is time-variation in the results (iid is violated), extremes are approached by "VaR which face time-variation in the return distribution incorporate volatility updating by ARCH or exponential smoothing models". In a more general approach it is possible to "allow for a time-varying conditional tail index"[16].This could be useful for dealing with Time-Varying Parameters (TVP)[17].

Peak Over Threshold (POT) or tail-fitting approach

In this variant[18] data is only considered when it surpasses a certain threshold u. The approach has been strongly developed for the interest of the insurance business, where only losses (pay outs) above a certain threshold are accessible[19]. One prescribes a high critical level u, and then studies each occasion the level is exceeded by the time series $(X_n), n \in N$. When an underlying distribution P belongs to the domain of max-attraction of a standard extreme value distribution, the distribution of an exceed value, that is, the conditional distribution of the height of $X_k - u$ given $X_k \geq u$, can be approximated up to a linear transformation by a generalized Pareto distribution.

$$H(x) = \begin{cases} 1-(1+\gamma x)_+^{-1/\gamma}, & \gamma \neq 0, \\ 1-e^{-x}, & \gamma \neq 0, \end{cases}$$

Such an approach allows us to consider not only a single absolute maximum (extreme) but to look at all "disadvantage" events, and even to consider them together with their respective heights. It is natural to suppose that, provided the threshold level u is high, the common distribution of the exceed points, as rare events, can be approximated by a distribution of a Poisson marked process. Many different questions could be considered in the EVT: Independent/Dependent Data; Stochastic Processes, Random Fields; Probabilities of Large Deviations; Exact Behaviour; Maxima and Excursions of Gaussian and Related Processes and Fields; Relationship between Continuous and Discrete Time.

3 Self-Organized Criticality

This approach is an established area in statistical physics. An extraordinarily wide range of physical and natural phenomena is power law distributed. The time series follow this kind of

[16] For details see: Wagner (2003).

[17] The basic EVT could be considered a particular case (the Time-Constant Parameter case) of the more general TVP approach.

[18] This variant is based on the second theorem in EVT; It was pionered by Pickands (1975), Balkema and de Haan (1974).

[19] See: Albeverio and Piterbarg (2006).

distribution. This has been found in a myriad of fields.[20] To explain this occurrence, Bak et al. (1987) introduced the concept of 'Self-Organized Criticality' (SOC), long-range spatio-temporal correlations that are signatures of deterministic chaos. The power law distribution is: $N(s) \approx s^{-\tau}$ where N is the number of observations at scale s and $\tau > 0$ is a parameter[21].

The tenet of this approach is that "events can and do happen on all scales, with no different mechanism needed to explain the rare large events, than that which explains the smaller, more common ones" (Bak, 1996)[22]. They can all be described as self-organizing systems that develop similar patterns over many scales. This is the consequence of the possible occurrence of coherent large-scale collective behaviours with a very rich structure, resulting from the repeated non-linear interactions among its constituents. For this, they "are not amenable to mathematical analytic descriptions and can be explored only by means of numerical experiments"[23], or simulations and often they are said to be computationally irreducible. In this case the traditional Decision Theory could not be applied any more.[24]

4 Differences Between SOC and TVP Literatures

Important differences between SOC and TVP literatures[25]: Classical extreme value theory is based on the assumption of i.i.d. observations that, vice versa, is typically violated. This is the case, for example, for return observations in the finance context[26].

The former exploits the data to find a distribution function whereas the latter generally uses models only for generating statistical signatures independently from actual data series.

More importantly, the former literature is more forecast oriented whereas SOC considers forecasting extreme events inherently infeasible, even if, there are now some trials concerning forecasting, through the study of the characteristic changes of the system before the crises.

Very often SOC properties are, with a few special exceptions, known only from simulation experiments while EVT literature is based on analytical proof.

Even if some authors model VaR-quantiles by an autoregressive conditional equation analogous to the way conditional variance is modeled in the prominent GARCH(1,1) setting, by performing numerical simulations and theoretical modeling, Lillo and Mantegna (2004) show that the non-linear behavior observed in real market crashes cannot be described by a GARCH(1,1) model.

They also show that the time evolution of the Value at Risk observed just after a major crash is described by a power-law function lacking a typical scale.

---

[20] From earthquakes (Hainzl et al. 1999) to music (Voss and Clark 1978) passing through wars (Roberts and Turcotte 1999), economics (Scheinkman and Woodford 1994) and forest fires (Drossel and Schwable 1992); Dessai S.; Walter M. E. (2000).

[21] Scott Moss (2002).

[22] For an introductory survey and explanation of SOC see Buckanan M. (2000).

[23] Sornette, D. (2002).

[24] Bankes (2002).

[25] Part of these ideas are taken from Scott Moss (2002) who discuss particularly TVP and SOC.

[26] Wagner N. (2003) concentrates on the issue of time-variation in the underlying return distribution.

5 Forecasting and SOC – Why Markets Crash

There are many examples of forecasting based on SOC ideas. An example of this approach could be Kaizoji who compares the statistical properties of index returns in a financial market, working an ensemble of stock prices, before and just after a major market crash. He found that the tail of the complementary cumulative distribution function of the ensemble of stock prices in the high value range is well described by a power-law distribution, $P(S > x) \approx x^{-\alpha}$, with an exponent that moves in the range of $1.09 < \alpha < 1.27$. Furthermore, he found that as the power-law exponents α approached unity, the internet bubble bursts. This suggests that Zipf's law for an ensemble of stock prices is a sign of bubble bursts.

If the system is a SOC, a different decision theory must be used. Following the Bankes (2002) approach, other kinds of decision theory different from the traditional one based on the optimality criterion could help to deal with the probability of extreme events, like that proposed starting from the results obtained by new Agent Based Modeling[27]. An "ensemble approach"[28] to prediction and not a single specimen data approach could also be used.

6 Conclusion

As a result, our short survey suggests that: (i) the results of pure statistical models could be erroneous when the system is a SOC; (ii) the forecasting limits of the SOC approach seem to have been removed; (iii) it seems time to explore more at large, extreme event causes and apply a different kind of decision theory.

References

[AS06] Albeverio S. and Piterbarg V.: Mathematical Methods and Concepts for the Analysis of Extreme Events in Nature and Society (The Frontiers Collection) S. Albeverio, V. Jentsch, H. Kantz (eds.), Springer (2006)

[BS93] Bankes, S.: Exploratory Modelling for Policy Analysis; Oper. Res. 41, 435–449 (1993)

[BS02] Bankes S.: Tools and techniques for developing policies for complex and uncertain systems; PNAS-May 14, 2002, vol. 99, suppl. 3, 7263–7266 (2002); www. pnas. org/cgi/doi/10.1073/pnas.092081399

[BV99] Bier, V.; Haimes, Y.; Lambert, J.; Matalas, N. and Zimmerman R.: A Survey of Approaches for Assessing and Managing the Risk of Extremes; Risk Analysis, Vol. 19, No. 1 (1999)

[BM00] Buckanan M.: Ubiquity: the science of history, Weidenfeld & Nicholson (2000)

[BR02] Buizza R.: Chaos and weather prediction January 2000; European Centre for Medium-Range Weather Meteorological Training Course Lecture Series ECMWF, 2002 (2002)

[BK75] Burry K. V.. Statistical Methods in Applied Science. John Wiley Sons (1975)

---

[27] Salzano (2006)

[28] This approach is largerly used for weather forecasts. See: Buizza (2002).

[DW00] Dessai, S. and M. E. Walter_ Self-Organised Criticality and the Atmospheric Sciences: selected review, new findings and future directions. Proceedings of the Conference on Extreme Events: Developing a Research Agenda for the 21st Century, Boulder, Colorado, US. (2000); http://www.tyndall.ac.uk/publications/briefing- notes/note11.pdf

[EH99] Einmahl H. J., de Haan L.: Empirical processes and statistics of extreme values, 1 and 2. AIO Course (1999), available at: www.few.eur.nl/few/people/Idehaan/aio/aiol.ps, www.few.eur.nl/few/people/ldehaan/aio/aio2.ps.

[FJ03] Feigenbaum J.: Financial physics, Rep. Prog. Phys. 66 1611-1649 PII: S0034-4885(03)25230-1 (2003), online at: stacks.iop.org/RoPP/66/1611 http://www.complexsystems.net.au/wiki/ Dynamics and statistics of multiscale systems

[KT06] Kaizoji T.: A Precursor of Market Crashes: Empirical laws of Japan's internet bubble; arXiv:physics/0510055 v4 – 17 Mar 2006 (2006)

[KS99] Kotz, S., Nadarajah S.: Extreme Value Distributions – Theory and Applications (1999); http://www.worldscibooks.com/mathematics/etextbook/p191/p191 chap1 1.pdf

[LM04] Lillo F., Mantegna R.: Dynamics of a financial market index after a crash; Physica A, Vol. 338, Issue 1–2. (2004)

[PR01] Pielke, R. A. Jr.: Room for doubt. Nature 410: 151 (2001)

[PR1a] Pielke, R. A. Jr.: The Role of Models in Prediction for Decision, Discussion Draft of 28 February – Paper prepared for Cary Conference IX: Understanding Ecosystems: The Role of Quantitative Models in Observations, Synthesis, and Prediction (2001a)

[SM06] Salzano M.: Una simulazione neo-keynesiana ad agenti eterogenei; in Terna and al. (eds.) Modelli per la complessità, Mulino (2006)

[SM02] Scott Moss: Policy analysis from first principles 2002;99;7267–7274 PNAS (2002); http://www.pnas.org/cgi/reprint/99/suppl3/7267.pdf

[SD02] Sornette, D.: Predictability of catastrophic events: Material rupture, earthquakes, turbulence, financial crashes, and human birth; http://www.pnas. org/cgi/content/full/99/suppl 1/2522 vol. 99 Suppl. 1; p. 2522–2529 (2002)

[SB02] Stewart T. R. and Bostrom A.: Extreme Events Decision Making – Workshop Report, Center for Policy Research – Rockefeller College of Public Affairs and Policy, University at Albany Decision Risk and Management Science Program National Science Foundation (2002); http://www.albany.edu/cpr/xedm/

[WA04] Willinger, W., Alderson, D., Doyle, J. C. and Li, L.: More "Normal" than Normal: Scaling Distributions and Complex Systems; Proceedings of the 2004 Winter Simulation Conference, R. G. Ingalls, M. D. Rossetti, J. S. Smith, and B. A. Peters, (eds.) (2004)

[WN03] Wagner N.: Estimating financial risk under time-varying extremal return behavior; OR Spectrum 25: 317–328 (2003)

Subject Index

Author Index

MIX
Papier aus verantwortungsvollen Quellen
Paper from responsible sources
FSC® C105338

If you have any concerns about our products,
you can contact us on
ProductSafety@springernature.com

In case Publisher is established outside the EU,
the EU authorized representative is:
Springer Nature Customer Service Center GmbH
Europaplatz 3, 69115 Heidelberg, Germany

Printed by Libri Plureos GmbH
in Hamburg, Germany